# CROPS RESIDUES AND BY-PRODUCTS IN ANIMAL FEEDING

# CROPS RESIDUES AND BY-PRODUCTS IN ANIMAL FEEDING

*Editor*
*Dr. Bibek Ghosh*

2007
**Gene-Tech Books**
**New Delhi - 110 002**

**ISBN 81-89729-67-5**

***Published by*** : **GENE-TECH BOOKS**
**4762-63/23, Ansari Road, Darya Ganj,**
**NEW DELHI - 110 002**
**Phone: 41562849**
**e-mail: genetechbooks@yahoo.co.in**

***Printed at*** : **Tarun Offset Printers**
**Delhi**

**PRINTED IN INDIA**

# Preface

Livestock play an important role in human society. Demand for meat and milk is soaring, and the world's livestock sector is growing at an unprecedented rate. This puts great pressure on the global natural resource base and, as demand continues to soar. In some countries, the productivity of the animals is very low, although experimental data suggest that their production potential is higher. The greatest constraint to livestock productivity is the shortage of feeds and forages. Some countries import cereal grains in an effort to increase livestock production and meet the ever-increasing demand for animal products. Higher livestock productivity, however, should be sought through better use of locally available feed resources. Crop residues, agro-industrial by-products and animal wastes, which are available in appreciable quantities, can play a significant role in the nutrition of ruminants.

Several of the crop residues and by-products available today are used commercially for purposes other than feeding. Examples are sugarcane bagasse used as fuel, straw used as mulch, bedding, fuel, building material or in the chemical industry. It is likely that technical developments will appear in the future which will make straw, bagasse, maize cobs etc. more attractive raw materials for industrial production. Such uses should be encouraged in areas where the crop residues and by-products cannot be utilised for animal feeding without long and expensive transportation.

The incentive to treat crop residues and agro-industrial by-products is less in areas where high quality forages and concentrate feeds of low cost are abundant than in countries where feed alternatives to straw are scarce. It may be argued that if by-products could replace part or most of the grass products in the diet, more land would be available for production of human food. In many countries where this would apply, the market for human products is saturated and the world market prices do not pay for the extra work.

The purpose of this book is to assist researchers in developing countries, to develop livestock feeding systems based on the available resources which are mainly crop residues, dry and/or mature pastures and agro-industrial by-products. It will of great interest to research scientists, extension workers, students, farmers and agricultural managers.

***Editor***

# Contents

# 1

# Potentials of Crop Residues as Animal Feed

The plant parts remaining in a field after the harvest of a crop, which include stalks, stems, leaves, roots, and weeds is known as crop residues. There are two types of agricultural crop residues: field residues and process residues.

Field residues are materials left in an agricultural field or orchard after the crop has been harvested. These residues include stalks and stubble (stems), leaves, and seed pods. Good management of field residues can increase efficiency of irrigation and control of erosion.

Process residues are those materials left after the processing of the crop into a usable resource. These residues include husks, seeds, bagasse, and roots. Crop residues can be used as animal fodder and soil amendment, and in manufacturing.

Asia is the leading region in the production of crop residues from wheat, rice, paddy, millet, buckwheat, sugarcane, nuts, oilplants and vegetables as well as fruit and berries. North and Central America is the leading region in the production of fibrous crop residues from maize, sorghum and pulses. Europe is leading in residue yields from barley, rye, mixed grain and roots and tubers. USSR leads in residue yield: from oats. In 1981, global production of fibrous crop residues from different crops in trillion tonnes was: maize 1.0, wheat, rice and paddy, pulses each 0.5, sorghum and oilplants each 0.3, barley and sugarcane each 0.2, roots and tubers 0.17, millet and nuts each 0.13, fruit and berries 0.11, vegetables 0.08, oats 0.06 and rye 0.05.

If all crop residues are counted, the world average would be 6166 Mcal ME and 3.4 tonnes of DM/LU of grass eater per year. The amount of crop residues in tonnes DM/LU of 'grasseaters/year was: 6.66 in North and Central America, 3.56 in Asia, 3.54 in Europe, 2.94 in USSR, 2.20 in Africa, 1.87 in South America and 1.06 in Oceania. Large difference were observed between individual countries. Much of these differences could be explained by climatic factors, level of technology in agricultural production systems, availability of natural pastures and forest lands for grazing, availability of drinking water, extent of livestock diseases, religious and cultural factors and political and economic regulation systems.

Yields of crop residues have risen proportionally more than livestock numbers from 1970 to 1981. The grain eater population in percent of total livestock population is the highest in Europe (33.7%) and the lowest in Oceania (3.3%). Countries which have high crop residues /LU of grass eaters tend to have also high grain eater numbers. It is evident that at present there are more fibrous crop residues available per LU of grass eaters than a decade ago. From an economic and feed security point of view, research should be strongly directed towards improving the rate of utilisation of fibrous crop residues as livestock feed.

**Crop Residues for Ruminant Livestock**

In some countries ruminant livestock is maintained mainly on pastures, forages and grains derived from arable land while in some other countries ruminants receive most of their energy needs from crop residues and/or natural pasturelands, meadows and wastelands. Studies on availability of crop residues as livestock feeds have been focused so far mainly on individual countries, groups of countries or regions.

In a study produced at Winrock International Livestock Research and Training Centre, Fitzhugh et al give global figures on various feed energy resources available for ruminants. Origins of the different feed resources and their estimated total energy value given in that study for 1970 and the projections for 2 000 are referred to in Table 1.

*Table 1. Feed energy resources available to ruminants in the whole world*

| *Land and type of feed* | *Year* | | *Energy source as % of world total* | |
|---|---|---|---|---|
| | *1970* | *2000* | *1970* | *2000* |
| | *ME billion Mcal* | | | |
| Permanent pasture, meadow, forage | 4 820 | 5 610 | 38 | 38 |
| Non-agricultural land forage | 1 019 | 719 | 8 | 5 |
| Arable land: | | | | |
| Forage | 3 115 | 3 775 | 25 | 26 |
| Crop residue | 2 945 | 3 560 | 23 | 24 |
| Grain | 439 | 767 | 4 | 5 |
| Oilseeds | 83 | 164 | 1 | 1 |
| Agro-industrial by-products | 128 | 215 | 1 | 1 |
| Total | 12 549 | 14 810 | 100 | 100 |

Table 1 shows that crop residues account for about 23 percent of all the feed energy resources available to ruminants in the world in 1970. Permanent pastures and meadows account for 38 percent, forages from arable land for 25 percent and agro-industrial by-products only for 1 percent.

The relative importance of various energy sources remains more or less the same in the projections made for the year 2000. In both developed as well as developing countries crop residues account for about 1/4 of the total feed energy suitable for ruminant livestock (Table 2).

*Table 2. Comparison of the feed energy resources in developed and developing countries in 1979*

| *Land and type of feed* | *Developed countries* | | *Developing countries* | |
|---|---|---|---|---|
| | *ME (bill Mcal)* | *%* | *ME (bill Mcal)* | *%* |
| *Permanent pasture,* meadow forage | 1 970 | 33.6 | 2 850 | 42.6 |
| Non-agricultural land forage | 266 | 4.5 | 753 | 11.3 |
| *Arable land:* | | | | |
| Forage | 1 720 | 29.4 | 1 395 | 20.9 |
| Crop residues | 1 365 | 23.3 | 1 580 | 23.6 |
| Grain408 | 7.0 | 31 | 0.5 | |
| Oilseeds | 69 | 1.2 | 14 | 0.2 |
| *Agro-industrial* | | | | |
| by-products | 61 | 1.0 | 67 | 1.0 |
| Total | 5 589 | 100 | 6 690 | 100 |

Compared to developing countries, in developed countries, more forages and grain are obtainable from arable land, while less forage is obtainable from natural pasturelands and meadows and nonagricultural land. According to Fitzhugh et al, forages constitute 84.5 percent of the metabolisable energy intake of cattle and buffaloes, 93.8 percent of that of sheep and goats, 17.7

percent of that of pigs, 8.6 percent of that of poultry and 93.3 percent of the ME intake of other animals.

### Fibrous Crop Residues

A global review on the annual yields of fibrous crop by-products was recently done at FAO headquarters. The study involved some 100 commodities, 212 countries and a period covering the years from 1970 to 1981. The results of the study have been published from the regional and socio-economic group point of view. It was not possible to handle individual countries in that connection. However the methodology applied is described in detail in that publication. The amounts of fibrous crop by-products, derived from cereals and other crops, in different regions in 1981 are shown in Fig. 1. It also gives the quantities of livestock units (LU) for grass eaters and grain eaters (pigs, poultry) separately for each region.

Asia is the leading region as far as the livestock numbers and production of fibrous crop residues are concerned. In North and Central America, Europe, and South America the size of livestock population is similar but North and Central America has a very high fibrous crop residue figure (Fig. 1). Quantitatively taken in Africa and USSR the fibrous crop residue situation is quite similar to each other. In Oceania, on the other hand, very little residues are produced. The amounts of fibrous crop residues produced per LU of grass eaters in different regions are given in Table 3.

The World's "grainstorehouse" North and Central America produced 6.66 tonnes of fibrous crop residues per LU of grass eaters in 1981. Europe and Asia ranked the second in this respect with some 3.5 tonnes/LU. Europe, however, has a very large grain eater population - 33.7 percent of the total LU numbers, compared to

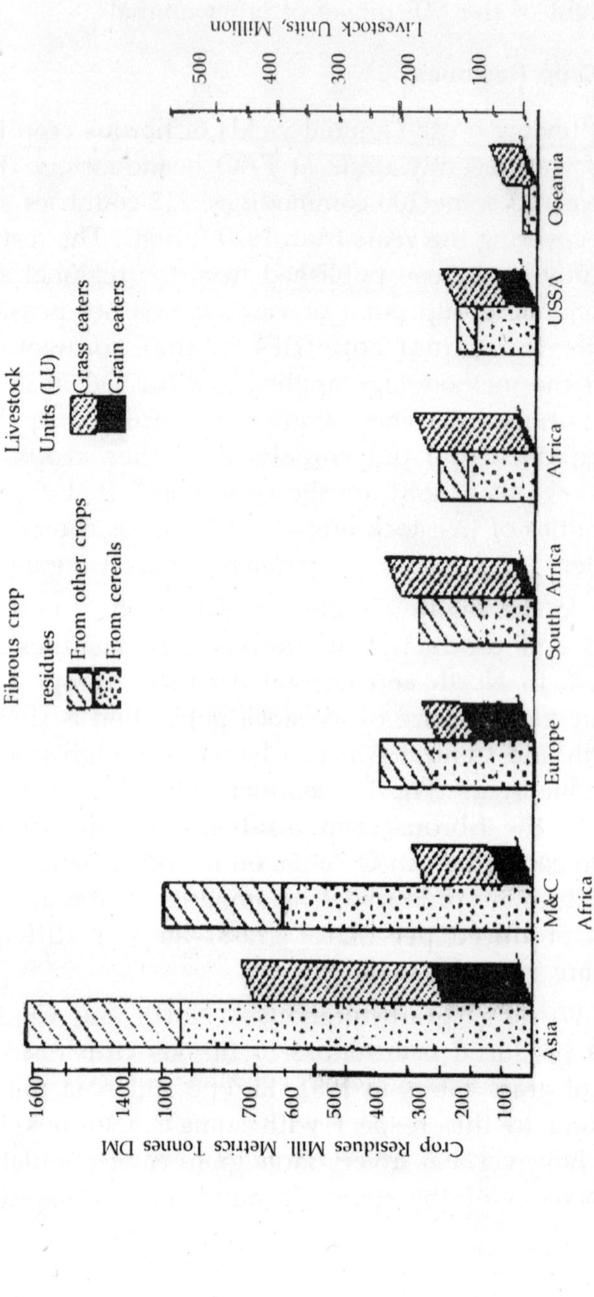


*Figure 1. Regional differences in the yield of fibrous crop residues from cereals and other crops, and in the numbers of grass eaters and grain eaters in 1981.*

Asia's 17.1 percent. USSR produced 2.94 tonnes of fibrous crop residues per LU of grass eaters. Like Europe, USSR also has a high grain eater population (26.1 percent of total LU number).

*Table 3. Total amount of fibrous crop residues in different regions in 1981*

| | *Fibrous crop residues produced in mill tonnes DM* | *Grass eaters in mill LU* | *Tonnes DM of residues/ LU of grass eaters* | *Percentage of grain eaters of total LU* |
|---|---|---|---|---|
| Asia | 1628 | 457 | 3.56 | 17.1 |
| N & C America | 1193 | 179 | 6.66 | 17.9 |
| Europe | 504 | 142 | 3.54 | 33.7 |
| S. America | 380 | 203 | 1.87 | 8.6 |
| Africa | 343 | 156 | 2.20 | 5.5 |
| USSR | 320 | 109 | 2.94 | 26.1 |
| Oceania | 53 | 50 | 1.06 | 3.3 |

In Africa and South America near 2 tonnes of fibrous crop residues were produced per LU of grass eaters. In both regions the number of grain eaters is very moderate (Table 3). In Oceania, which is a typical area for grazing ruminants, only 1 tonnes of fibrous crop residues was produced per LU of grass eaters in 1981. The grain eater population was only 3.3 percent of total LU in Oceania.

North and Central America is the leading region in the production of fibrous byproducts from maize, sorghum and pulses. Asia is the leading region in the production of residues from wheat, rice and paddy, millet, buckwheat, sugarcane, nuts, oilplants, vegetables as well as fruit and berries. Europe leads the residue fields from barley, rye, mixed grain and roots. USSR leads in residue yields from cats. Referring to the world's total crop residue figures maize yielded the largest amount of fibrous crop residues (1.0 trillion tons).

Wheat, rice and paddy and pulses, each yielded about half of the amount compared to maize (0.5 trillion tonnes), sorghum and oilplants each yielded 0.3 trillion tons, barley and sugarcane each yielded 0.2 trillion tons, roots and tubers 0.17 trillion tons, millet and nuts, each about 0.13 trillion tons, fruit and berries about 0.11, vegetables 0.08, oats 0.06 and rye yielded 0.05 trillion tons of fibrous residues. If all given crop residues are counted this would give 3.4 tonnes and 6166 Meal ME/LU of grass eaters per year on an average in the whole world.

## Crop By-product Situation in Different Regions

In 1981, Africa carried about 12 percent of the world's grass eater population and it produced some 343 million metric tonnes of DM from crop residues which accounts for 7.7 percent of the world total. 68.8 percent of the African crop residues were derived from cereals, 12 percent from oilplants and pulses, 4.9 percent from sugarcane etc. Africa has clearly defined areas with rather definite preferences for certain cereals. Maize is prevalent in South-East Africa, in Egypt and in a few South-Western countries.

Millet and sorghum are prevalent in Middle and West Africa. Wheat is popular in North Africa. Rice is preferred in several countries on the South West coast and in the East Coast Islands. Two countries, i.e. Reunion and Mauritius are specialised primarily in sugarcane production. There is a definite belt of countries beginning from the South East coast and extending to the South West African coast line, which have produced over 2 tonnes/LU of grass eaters of fibrous crop residues.

On the other hand, North African countries as well as Botswana and Namibia in the South had very low crop residue figures. In 1981, South America produced 8.6

percent of fibrous crop residues and it carried about 15.7 percent of the grass eater population of the world total figure. 47 percent of the fibrous crop residues were derived from cereals, some 14 percent from sugarcane, 20 percent from pulses and the rest from other sources. In most of the South American countries, maize was the main cereal producing fibrous crop residues. However, in Suriname and Guyana, the main cereal was rice and in Venezuela it was sorghum.

Several South American countries produced considerable quantities of fibrous crop residues from sugarcane. The total quantities of fibrous crop residues/ LU of grass eaters was less than one tonne in Bolivia, Uruguay, Venezuela, and Colombia. On the contrary, high figures were observable in Guyana (5.38 tonnes/LU) and Suriname (10.8 tonnes/LU) (Table 4).

The share of grain eaters in percentage of total LU number was particularly low in Argentina (2.6 percent), Colombia (3.5 percent) and Uruguay (1.7 percent) but rather high in Ecuador (23.9 percent), Guyana (42.2 percent), Suriname 32.7 percent and French Guyana (30 percent) (Table 4).

In 1981 Asia carried about 35.2 percent of the grass eaters of the world. In the same year 1 628 882 thousand metric tonnes of fibrous crop residues were produced in Asia, which represents 36.8 percent of the world total. Sixty-eight percent of these residues were derived from cereals, 14.5 percent from oilplants and pulses, 4.5 percent from sugarcane. The most important cereal crop was rice and paddy, the next in order were maize and wheat. The average amount of crop residues per LU of grass was 3.56 tonnes per year in 1981. Generally speaking Near and Far Eastern countries had more crop residues available per LU of grass eaters compared to Middle East figures.

*Table 4. Amounts of fibrous crop residues produced in South America in 1981.*

| | Crop residues '000 MD DM | Main crops | Grass eaters '000 LU | Grain eaters % total LU | Production of fibrous crop residues MT DM/LU grass eaters |
|---|---|---|---|---|---|
| Brazil | 209 678 | maize<br>cane<br>pulses | 86 275 | 12.4 | 2.43 |
| Argentina | 113 011 | maize<br>sorghum | 49 826 | 2.6 | 2.27 |
| Colombia | 18 302 | maize<br>cane | 22 437 | 3.5 | 0.81 |
| Paraguay | 7 132 | maize<br>pulses | 4 736 | 8.3 | 1.50 |
| Peru | 6 964 | maize<br>cane | 5 933 | 12.9 | 1.17 |
| Venezuela | 6 781 | sorghum<br>maize | 9 752 | 9.2 | 0.70 |
| Ecuador | 5 481 | maize<br>cane | 3 288 | 23.9 | 1.67 |
| Chile | 5 173 | maize | 4 137 | 11.4 | 1.25 |
| Uruguay | 2 961 | maize<br>wheat<br>rice<br>sorghum | 11 355 | 1.7 | 0.26 |
| Bolivia | 2 893 | maize<br>cane | 5 444 | 7.1 | 0.53 |
| Guyana | 1 378 | rice<br>cane | 256 | 42.2 | 5.38 |
| Suriname | 400 | rice | 37 | 32.7 | 10.80 |
| French Guyana | 9 | maize | 6 | 30.0 | 1.50 |

The average grain eater population is very modest in Asia (Table 3). However, there are great differences between areas and countries in this respect. For instance, the majority of the Middle Eastern countries have less than 5 percent of grain eaters while most of the Far Eastern countries have a large proportion of grain eaters (45 to 70 percent).

In some special cases (Hong Kong and Singapore) there are 97 percent of grain eaters. Moreover, several Near Eastern countries (Jordan, Israel, Lebanon, Cyprus) have a considerable grain eater population in relation to the total livestock population. A high proportion of grain eaters is often associated with a high availability of fibrous crop residues. Since Asia is a very large region, it is understandable that extremes in climatic and soil conditions are found in that area, which greatly influence crop and livestock production and are reflected in the figures obtained for different countries.

USSR carried about 9.4 percent of the total livestock units and produced 7.2 percent of the fibrous crop residues of the world in 1981. Crop residues accounted for about 2.94 tonnes per LU of grass eaters (Table 3).

In 1981, Oceania yielded 1.2 percent of the fibrous crop residues while they had 3.9 percent of grass eaters of the world total. 41 percent of the fibrous crop residues were derived from wheat, 13.7 from cane, and some 9 percent from barley, sorghum and nuts each. 83.7 percent of the crop residues came from Australia and the rest from the other islands of Oceania (Table 5). The quantity of residues per LU of grass eaters was 1.95 tonnes.

Nuts were the main crop in most of the small islands. The grass eater production systems depend heavily on grass in this region. The grain eater population as a percentage of total livestock quantity is very high in some of the small islands while the average for the whole region is very low (Tables 3 and 5). Also the quantity of fibrous by-products per LU of grass eaters is high in some islands. Some of the smaller islands belonging to Oceania were left out because there was not enough information available.

*Table 5. Amounts of fibrous crop residues produced in Oceania in 1981.*

| Country | Crop residues '000 tonnes DM | Main crops | Grass eaters '000 LU | Grain eaters % total LU | Production of fibrous crop residues MT DM/LU grass eaters |
|---|---|---|---|---|---|
| Australia | 44 377 | wheat | 35 445 | 3.3 | 1.25 |
| Papua-New Guinea | 2 425 | nuts | 108 | 71.4 | 22.5 |
| New Zealand | 2 119 | maize | 14 496 | 1.2 | 0.15 |
| Fiji | 1 500 | cane | 168 | 11.1 | 8.9 |
| Vanuatu | 493 | nuts | 59 | - | 17.6 |
| Samoa | 427 | nuts | 32 | 38.0 | 13.3 |
| Tonga | 270 | nuts | 25 | 41.5 | 10.8 |
| FrenchPolynesia | 257 | nuts | 10 | 60.9 | 25.7 |
| New Caledonia | 51 | nuts | 100 | 6.5 | 0.51 |
| Cook Islands | 33 | nuts | 8 | 27.3 | 4.1 |

In 1981, North and Central America produced about 27 percent of fibrous crop residues and had about 13.8 percent of grass eaters of the world total figure. 96 percent of the crop residues were produced by USA, Canada and Mexico. In this region 68.6 percent of crop residues were derived from cereal crops (40% from maize), 20 percent from pulses, 3.5 percent from sugar cane and the rest from other sources. Maize was the main residue producing cereal crop in USA, Mexico, El Salvador and Guatemala, wheat in Canada and sugarcane in most of the Caribbean Islands and in the Central American mainland (Table 6).

*Table 6. Amounts of fibrous crop residues produced in North and Central America in 1981*

| Country | Crop residues '000 tonnes DM | Main crops | Grass eaters '000 LU | Grain eaters % total LU | Production of fibrous crop residues MT DM/LU grass eaters |
|---|---|---|---|---|---|
| USA | 947051 | maize pulses | 114354 | 18.3 | 8.28 |
| Mexico | 102918 | maize | 30864 | 12.5 | 3.46 |
| Canada | 92546 | wheat | 11651 | 27.7 | 7.94 |
| Cuba | 18477 | cane | 6302 | 13.3 | 2.93 |
| Guatemala | 5751 | maize | 1792 | 20.6 | 3.20 |
| Dominican Rep. | 4439 | cane | 2362 | 6.1 | 1.88 |
| El Salvador | 2800 | maize | 1203 | 13.8 | 2.32 |
| Haiti | 2797 | cane | 1722 | 19.4 | 1.62 |
| Honduras | 2608 | cane | 2338 | 9.6 | 1.12 |
| Nicaragua | 2465 | cane | 2392 | 8.6 | 1.03 |
| Costa Rica | 1984 | cane | 2166 | 6.6 | 0.92 |
| Panama | 1633 | cane | 1615 | 8.0 | 1.01 |
| Jamaica | 1207 | cane | 337 | 29.9 | 3.58 |
| Puerto Rico | 659 | cane | 316 | 28.8 | 2.08 |
| Trinidad & Tobago | 542 | cane | 88 | 60.1 | 6.16 |
| Belize | 336 | cane | 55 | 16.7 | 6.11 |
| Guadelopue | 276 | cane | 88 | 25.0 | 3.14 |
| Barbados | 257 | cane | 30 | 49.1 | 8.57 |
| Martinique | 135 | fruit & berries | 60 | 39.4 | 2.25 |
| St Lucia | 122 | nuts | 14 | 30.0 | 8.71 |
| Dominica | 76 | nuts | 4 | 44.4 | 19.00 |

In this region 6.66 tonnes DM was yielded per LU of grass eaters on an average. Cane has very high potential for production of fibrous byproducts.

*Table 7. Amounts of fibrous crop residues produced in Europe in 1981*

| Country | Crop residues '000 tonnes DM | Main crops | Grass eaters '000 LU | Grain eaters % total LU | Production of fibrous crop residues MT DM/LU grass eaters |
|---|---|---|---|---|---|
| France | 79911 | wheat | 23095 | 23.3 | 3.46 |
| Italy | 55928 | maize | 9529 | 31.2 | 5.87 |
| Poland | 49334 | rye | 12828 | 34.6 | 3.84 |
| Romania | 40531 | maize | 8281 | 37.9 | 4.92 |
| Germ.Fed.Rep | 38692 | barley | 14078 | 36.6 | 2.75 |
| Spain | 34275 | oilplants | 6494 | 38.2 | 5.28 |
| Yugoslavia | 31929 | maize | 6422 | 35.1 | 4.97 |
| UK | 27787 | barley | 15493 | 22.7 | 1.79 |
| Hungary | 26580 | maize | 2188 | 61.6 | 12.15 |
| German Dem. Rep. | 17365 | barley | 5447 | 45.9 | 3.19 |
| Greece | 16360 | oilplants | 2563 | 22.8 | 6.38 |
| Czechoslovakia | 15775 | wheat | 4652 | 40.0 | 3.39 |
| Belgium | 15545 | maize | 3247 | 35.4 | 4.79 |
| Denmark | 11006 | barley | 2706 | 54.2 | 4.07 |
| Sweden | 9718 | barley | 1848 | 35.6 | 5.26 |
| Austria | 7964 | maize | 2330 | 36.4 | 3.42 |
| Netherlands | 5060 | roots | 4743 | 48.1 | 1.07 |
| Belgium-Luxemb. | 4585 | roots+tub. | 2853 | 40.8 | 1.61 |
| Portugal | 4131 | oilplants | 1880 | 40.7 | 2.20 |
| Finland | 3633 | barley | 1634 | 25.8 | 2.22 |
| Ireland | 2715 | barley | 6500 | 6.8 | 0.42 |
| Albania | 1918 | wheat | 728 | 9.4 | 2.63 |
| Switzerland | 1870 | roots | 1850 | 27.8 | 1.01 |
| Norway | 1632 | barley | 1177 | 18.2 | 1.38 |
| Malta | 29 | vegetable | 17 | 52.8 | 1.71 |
| Lichtenstein | 2 | roots | 5 | 28.6 | 0.40 |
| Iceland | 2 | roots | 149 | 3.9 | 0.01 |
| Faeroe Islands | 1 | roots | 10 | 0.0 | 0.10 |

Canada and many of the Caribbean Islands have a rather high grain eater population in percent of the total LU number. European countries produced 11.4 percent of fibrous crop residues and carried 11 percent of grass eater population of the world total. 67 percent of the crop residues were derived from cereals, 11.2 percent from roots and tubers, and 10.7 percent from oilplants. Crop residues were derived mainly from barely and oats in North Europe, from rye in Poland, from wheat and maize in Middle Europe and in the South where also oilplants play a significant role. In the Northern Islands crop residues are derived mainly from roots and vegetables and yield per LU is very low.

The average figure for Europe was 3.54 tonnes DM/ LU of grass eaters in 1981. In 78 percent of the European countries this figure fits between 1 and 6 (Table 7) while for instance in Asia only 45 percent of the countries fit within this range. Europe has a very high grain eater population compared to other regions (Table 3). The percentage of grain eaters in percent of total LU population, however, fits between 20 and 60 in 78 percent of European countries.

**Potential Supply of Crop Residues and Agro-Industrial By-Products in Southeast Asia**

Table 8 presents the major crop production in five Southeast Asian countries during 1980/1981 season. Data indicate that the main food crops of the region are rice, sugar-cane, cassava, maize and other plantation crops such as coconut and oil palm. The potential supply of their residues and agro-industrial by-products are given in Tables 9 and 10. It is worth noting that these crop residues and agro-industrial by-products are not totally available for feeding of livestock for three reasons.

*Table 8. Production of Major Field and Plantation Crops in Some Selected Southeast Asian Countries*

| | Production by country | | | | | |
|---|---|---|---|---|---|---|
| | Indonesia | Malaysia | Philipines | Thailand | Vietnam | Asia |
| | '000 metric tons | | | | | |
| Rice (Oryza sativa) | 33 000 | 2 147 | 7 720 | 19 000 | 12 570 | 376 232 |
| Maize (Zea mays) | 3 991 | 8 | 3 176 | 3 700 | 540 | 86 570 |
| Sorghum | - | - | - | 380 | 35 | 20, 566 |
| Sweet Potatoes | 2 079 | 37 | 1 100 | 348 | 2 400 | 137 108 |
| Cassava (Manihot spp.) | 13 726 | 360 | 2 300 | 17 900 | 3 400 | 47 584 |
| Banana (Musa spp.) | 1 556 | 375 | 893 | 2 021 | 900 | 14 788 |
| Pineapple (Ananas comosus) | 266 | 207 | 1 200 | 1 800 | 350 | 5 003 |
| Sugar cane (Saccharum officinarum) | 17 560 | 850 | 20 450 | 18 600 | 3 900 | 290 357 |
| Soyabean (Glycine max) | 653 | - | 8 | 120 | 64 | 10 320 |
| Groundnuts (Arachis hypogaea) | 855 | 23 | 50 | 112 | 80 | 11 408 |
| Cotton seed (Gossypium spp.) | 7 | - | 3 | 153 | 3 | 11 871 |
| Coconut (Cocos nucifera) | 10 800 | 1 207 | 10 050 | 900 | 350 | 30 803 |
| Oilpalm kernels (Elaeis quineensis) | 131 | 588 | 2.5 | 2 | - | 769 500 |
| Rubber (Hevea brasiliensis) | 937 | 1 590 | 65 | 510 | 48 | 3 569 |

Firstly, the extraction rate which has been calculated from the primary products as given by Devendra is a general average which in turn is subject to error due to variation in harvesting and handling methods from place to place.

Secondly, accurate statistics on processing of commodities especially home processing is not complete thus making the theoretical estimation very inaccurate.

Finally, the potential availability may differ greatly from what is or could actually be used for feeding due to their value in non-feed utilisation.

Castillo has comprehensively discussed six broad areas in utilisation of fibrous residues in the Asian context. These include their utilisation as animal feed, as fertiliser or soil conditioner, as fuel or sources of energy, as a source of building/construction material, as sources of herbal medicines, biologics or chemicals and for sociological or cultural applications.

*Table 9. Potential Supply of Crop Residues in Some Selected Southeast Asian Countries*

| | *Production by country* | | | | | |
|---|---|---|---|---|---|---|
| | *Indonesia* | *Malaysia* | *Philipines* | *Thailand* | *Vietnam* | *Total* |
| | *'000 metric tons* | | | | | |
| Rice straw | 36 300 | 2 362 | 8 492 | 20 900 | 13 827 | 81 881.0 |
| Maize stover | 3 991 | 8 | 3 176 | 3 700 | 540 | 11 415.0 |
| Maize husk | 559 | 1.1 | 445 | 518 | 75.6 | 1 598.7 |
| Sorghum stover | - | - | - | 760 | 70 | 830.0 |
| Sorghum head w/o grain | - | - | - | 152 | 14 | 166.0 |
| Sweet potato vine | 624 | 11 | 330 | 104 | 720 | 1 789.0 |
| Cassava leaves | 1 098 | 29 | 184 | 1 432 | 272 | 3 015.0 |
| Banana stem and leaves | 3 423 | 825 | 1 965 | 4 446 | 1 980 | 12 639.0 |
| Banana fruit wastes | 467 | 112 | 268 | 606 | 270 | 1 723.0 |
| Pineapple wastes | 186 | 145 | 840 | 1 260 | 245 | 2 676 |
| Sugar-cane tops and leaves | 5 268 | 255 | 6 135 | 5 580 | 1 170 | 18 408.0 |
| Soyabean stover and pods | 1 306 | - | 16 | 240 | 128 | 1 690.0 |
| Groundnut vine | 419 | 11.3 | 24.5 | 54.9 | 39.2 | 548.9 |

*Table 10. Potential Supply of Agro-Industrial By-Products in Some Selected Southeast Asian Countries*

| | Production by country | | | | | |
|---|---|---|---|---|---|---|
| | Indonesia | Malaysia | Philipines | Thailand | Vietnam | Total |
| Concentrates | '000 metric tons | | | | | |
| Rice bran | 3 300 | 215 | 772 | 1 900 | 1 257 | 7 444.0 |
| Broken rice | 1 650 | 107 | 386 | 950 | 629 | 3 722.0 |
| Maize bran | 360 | 0.7 | 286 | 333 | 48.6 | 1 028.3 |
| Maize germ meal | 678 | 1.4 | 540 | 629 | 91.8 | 1 940.2 |
| Cassava wastes | 7 824 | 205 | 1 311 | 10 203 | 1 938 | 21 481.0 |
| Sugar-cane molasses | 702 | 34 | 818 | 744 | 156 | 2 454.0 |
| Soyabean meal | 475 | - | 6 | 87 | 46.5 | 614.5 |
| Groudnut meal | 462 | 12.4 | 27 | 60 | 43.2 | 604.6 |
| Cotton seed meal | 3.2 | - | 1.3 | 69 | 1.4 | 74.9 |
| Copra meal | 4 320 | 483 | 4 420 | 360 | 140 | 9 723.0 |
| Palm kernel meal | 29 | 129 | 0.6 | 0.4 | - | 159.0 |
| Rubber seed meal | 515 | 875 | 36 | 280 | 26.4 | 1 732.4 |
| Roughages | | | | | | |
| Rice hulls | 5 280 | 344 | 1 235 | 3 040 | 2 011.2 | 11910.2 |
| Maize cob | 798 | 1.6 | 635 | 740 | 108 | 2 282.6 |
| Pineapple bran | 10.6 | 8.3 | 48 | 72 | 14 | 152.9 |
| Palm press fiber | 484 | 1871.0 | 8 | 6.7 | - | 2 389.7 |
| Sugar-cane bagasse | 3 863 | 187 | 4 499 | 4 092 | 858 | 13 499.0 |

## Role of Crop Residues and Agro-Industrial By-products

Despite the fact that crop residues and agro-industrial by-products are currently used for various purposes, their importance as feedstuffs particularly for ruminants in small farm systems cannot be over-emphasised. With fewer fibrous agro-industrial by-products which are conventionally used in non-ruminant feeding, opportunities for offering them to ruminants or for improving their feeding value appear to be smaller and smaller due to rapid advances in nutritional sciences. Better utilisation of these less fibrous feedstuffs for non-ruminant feeding in Southeast Asia can be obtained by increasing local production of protein feedstuff in order to keep up with the regional deficiency.

Khajarern and Khajarern reported that the deficiency of protein concentrates in Southeast Asia was 3.9 million tons or 62.4% of the requirement. Exploitation for novel protein feeds consequently appears to be of the first priority for animal nutritionists in this region. On the other hand, fibrous crop residues seem to be currently utilised at lower efficiency levels even though they have been used as the main forage source for ruminants in the region for hundreds of years. Among the fibrous crop residues potentially available in Southeast Asia, rice straw, sugar-cane tops and leaves, banana residues and maise stover are of primary importance whereas other crop wastes are of secondary importance (Table 9).

For feeding purposes, rice straw is commonly used as main roughage, during the dry season, while other crop residues are rarely fed to livestock in Southeast Asian small farms. Traditionally, rice straw is stored, at the time of harvest, near the farmer's house either in the form of a stack or in a thatched hut. During the forage shortage, straw is fed to livestock either alone or mixed with a small supplement of green forage and generally without any concentrates. It has long been established that rice straw has low protein (3–5%), calcium (0.25- 0.55%), phosphorus (0.02–0.16%) but is high in crude fiber (26-34%) and silica (12–16%) with slightly lower lignin than other cereal straws.

The TDN for cattle is relatively low (35.9 to 50.3%). Furthermore, the amount an animal can eat is limited to less than 3 percent of body weight due to the slow rate at which it is fermented in the rumen resulting in relatively low and variable digestibility. In short, rice straw is a poor roughage which generally provides little or no surplus energy over maintenance requirement. Jackson concluded in his review that the only way to obtain more animal products from straw in the Asian setting was to improve digestibility and intake so that more energy was available for productive purposes.

Research aiming for better utilisation of the main crop residues in Southeast Asia, namely rice straw, has been undertaken by two major means. One is by improving the utilisation of untreated straw by various supplementations including energy, protein/nonprotein nitrogen and/or minerals, whereas the other is by treatments of the residue prior to feeding with or without further supplementation. Both of these means, however, have not created appreciable impact for application in the small farm levels at present. This consequently suggests that much more work has to be done not only in the technical but also in the extension and sociological aspects.

The utilisation of rice straw as a significant proportion of the ration would only be successful if the limitations of low crude protein and poor digestibility could be overcome. Supplementation with energy and protein sources is a prerequisite to increasing the efficiency of utilisation. Energy sources such as molasses, cassava chips and sweet potato chips are appropriate for this purpose, while rice bran, copra cake, palm kernel cake, soybean meal, groundnut meal, fish meal, urea and other green leaves can supply protein to the level that can support the optimum dry matter intake. O'Donovan and Chen, feeding untreated chopped straw (25 or 35%) with 25, 33 or 45 percent of molasses along with sweet potato chip-soybean meal-urea, obtained 0.46–0.82 kg body weight gain/day, in growing hiefers.

Devendra supplemented molasses at 40 percent of ration containing 40 percent rice straw and noticed an increase in dry matter intake to 3.3% of sheep body weight. He also noted that maximum dry matter and organic matter digestibility were obtained at 20 percent rice straw with 60 percent molasses. At this level, rice straw intake and utilisation were significantly increased. In a subsequent study, Devendra reported that when urea molasses was added to rice straw diets having maize as the main energy source, nitrogen intake and retention were significantly increased.

The favourable effect of molasses on maize is not only due to increased palatability but it is also an ideal source for urea utilisation in rice straw diets. With water buffalo, Suriyajantratong et al supplemented 10 percent molasses with either 0, 1, 2, 3 or 4 percent urea to rice straw diets and observed a progressively reduced weight loss as the urea level increased. The molasses supplementation without urea, however, produced higher body weight loss than in the animals on rice straw. Rice straw consumption in the former group was significantly reduced.

Similarly, Wanapat et al. noted the decreases in rice straw intake, body weight gain and dry matter digestibility when cattle on a rice straw diet were supplied with 2 kg/hd/d of cassava chip. They explained that this adverse effect was due to the starchy nature of cassava chip which brought about the equivalent decrease of roughage intake and parakeratosis. Devendra emphasised the importance of maintaining a balance between carbohydrate sources such as molasses with rice straw and dietary nitrogen if the maximum utilisation of rice straw is to be obtained.

Any diets containing less than 20 percent of rice straw supplemented with high levels of energy and protein sources would reduce the roughage utilisation. Another way to increase the utilisation of untreated rice straw with low inputs of concentrates is to include in the diets forages with a relatively high crude protein content. Positive dry matter digestibility, crude protein digestibility and nitrogen retention in sheep being fed rice straw, 10–60 percent DMI of leucaena and minerals were observed by Devendra. Similar responses were observed, in cattle receiving up to 50 percent DMI of leucaena, by Intramongkol et al and in cattle and water buffalo receiving rice straw rations with 2 kg/hd/d concentrates containing poultry manure up to 25 percent.

Other high protein forages have also been demonstrated to be suitable supplements in rice straw rations. These include kenaf (*Hibiscus* spp.) leaves, pigeon pea (Cajanus cajan) straw and stylosanthes, cassava leaves, and gliricidia leaves. Devendra concluded in his review on the utilisation of untreated straw that the optimum level of rice straw in a diet is about 20 percent and some of the aforementioned forages have potential value in increasing the utilisation of straw. Further, economic justification between straw treatment to improve its nutritive value and the low cost, untreated straw-based feeding system that can combine various crop residues and non-conventional feed resources has been questioned.

# 2

# Crop Residues and Agro-industrial By-products

Doubts about the usefulness of feeding standards for ruminants in developing countries began to be manifested at the time that the first feeding systems were being established in Cuba using non-conventional feed resources such as molasses. The issue was further emphasised as the results of experiences in Asia with alkali-treated straws. The significant features of these developments were not the feeding of molasses or the alkali-treated straw, since both these feeds had been in use for decades in the industrialised countries, without giving cause for doubting the nutritive value of these ingredients, as assessed by conventional methods.

The difference was in the magnitude of the contribution of the molasses and the straw to the animals' diet. This rarely exceeded 10–30 percent in the industrialised countries as there was ample availability of high quality feeds such as cereal grain and protein-rich meals to complete the diet. In contrast, the feeding regimes planned around molasses and straw in developing countries envisaged these ingredients supplying the major part of the ration, due to physical, financial and political constraints attached to the use of cereal grains and protein meals.

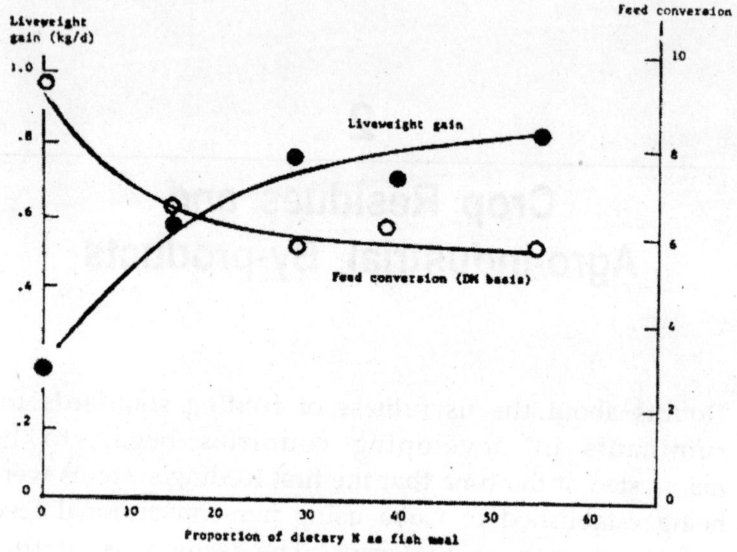


*Figure 1. Small amounts of fish meal improve the performance of steers fed a basal diet of molasses/urea*

One characteristic common to both molasses and to alkali-treated straw is the magnitude of the response to small amounts of proteins of high biological value when this is in a form which can escape the rumen fermentation. (Figures 1 and 2) These responses are much greater than is observed when comparable amounts of fish meal are added to a cereal grain-based diet. The second common feature of molasses and alkali-treated straw is that even when supplemented with fish meal, the efficiency of utilisation of the digested energy is considerably less than what is observed when the basal diet contains appreciable amounts of cereal grain. This "inefficiency" is particularly noteworthy when the host animal has a high glucose demand.

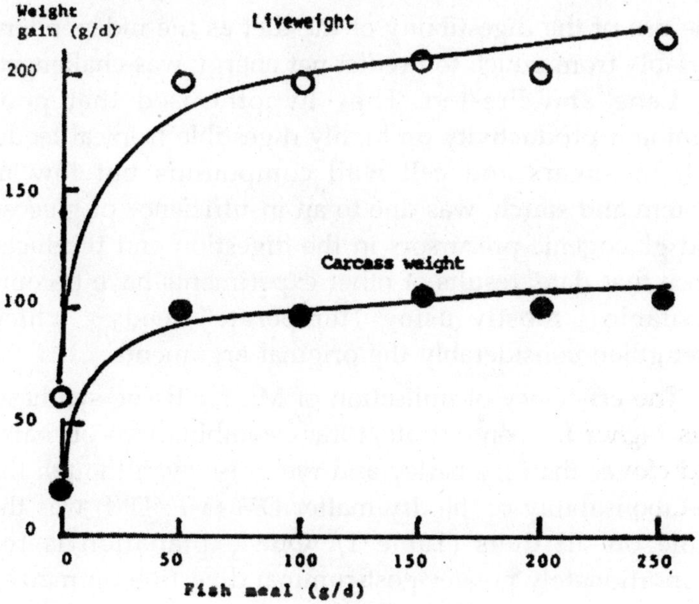


*Figure 2. Small amounts (50 g/d) of fish meal increase dramatically the performance of cattle fed ammonia (from urea)-treated rice straw in Bangladesh*

What distinguishes molasses and alkali-treated straw from cereal grains is the complete absence of starch, the negligible content of protein, and relatively high concentrations of soluble sugar. In other words, the nutrients in molasses and in alkali-treated straw are only digested in the rumen; in contrast, the nutrients in cereal grains can be digested either in the rumen or in the intestine. Many of the reasons for the failure of feeding standards to predict attainable levels of performance on molasses and straw-based diets can be explained by interactions and associated effects among nutrients and between nutrients and their sites of digestion.

## Digestibility of a Feed

The use of the digestibility of the diet as the independent variable from which to predict net energy was challenged by Leng and Preston. They hypothesised that poor ruminant productivity on highly digestible tropical feeds, rich in sugars and cell wall compounds but low in protein and starch, was due to an insufficiency of glucose and glucogenic precursors in the digestion end products. Since that date, results of other experiments have become available - mostly using "temperate" feeds - which strengthen considerably the original arguments.

The efficiency of utilisation of ME for tissue synthesis was higher for concentrate/forage combinations of maize and clover than for barley and rye grass even though the metabolisability of the dry matter DM (ME/DM) was the same on all diets (Table 1). One explanation is the proportionately greater post-ruminal digestion on maize/clover compared with barley/grass.

*Table 1. Efficiency of utilisation of metabolisable energy for fattening in cattle is greater for combinations of feeds containing maize or clover than for barley and ryegrass. The ME concentration was similar for all ration combinations.*

| | Efficiency ($k_f$= %) |
|---|---|
| Ryegrass and barley | 36 |
| Ryegrass adn maize | 42 |
| Clover and barley | 44 |
| Clover and maize | 50 |

Work in Australia (Table 2) showed that pangola grass was used more efficiently (28 %) than setaria grass (17 %) for tissue synthesis in cattle although both had the same digestibility and were fed at the same rates; the authors mention the superior glucogenic potential of pangola as one possible explanation for the difference. Rice polishings were better than cassava root meal for supplementing sugar cane. The starch in rice polishings escapes rumen fermentation almost quantitatively whereas the starch of cassava root meal is fermented rapidly in the rumen.

*Table 2. Efficiency of utilisation of metabolisable energy for fattening cattle is greater for pangola grass than for setaria even though the concentration of ME was the same in both feeds*

| | Concentration of ME in DM MJ/kg DM | Efficiency ($k_f$ = %) |
|---|---|---|
| Setaria grass | 8.0 | 17 |
| Pangola grass | 8.1 | 28 |

Glucose entry rates were higher when rice polishings rather than cassava root meal was the supplement in sugar cane diets. Supplementary energy as maize grain improved feed conversion efficiency in cattle fed sugar cane whereas the same amount of molasses energy depressed feed efficiency. Cattle nourished by rumen infusion of VFA and abomasal infusion of casein increased their nitrogen retention as the proportion of propionic acid in the infused VFA was increased (Figure 3).

Convincing evidence concerning the need for glucogenic compounds in the end-products of digestion was provided by Tyrrell et al (Figure 4). Acetic acid infused into the rumen of animals receiving a basal diet of low glucogenic potential was less efficiently utilised for tissue synthesis than when the infusion was given with a basal diet of high potential for providing glucose.

The superiority of propionic acid for fat synthesis compared with acetic acid, observed in the original work of Armstrong and Blaxter and Armstrong et al and the absence of differences between these two fatty acids in the experiments of Orskov and Allen can also be explained in terms of the glucogenic potential of the basic diet. The diet used by Armstrong et al was dried grass whereas Orskov and Allen gave the different VFA mixtures to animals fed mainly on barley grain.

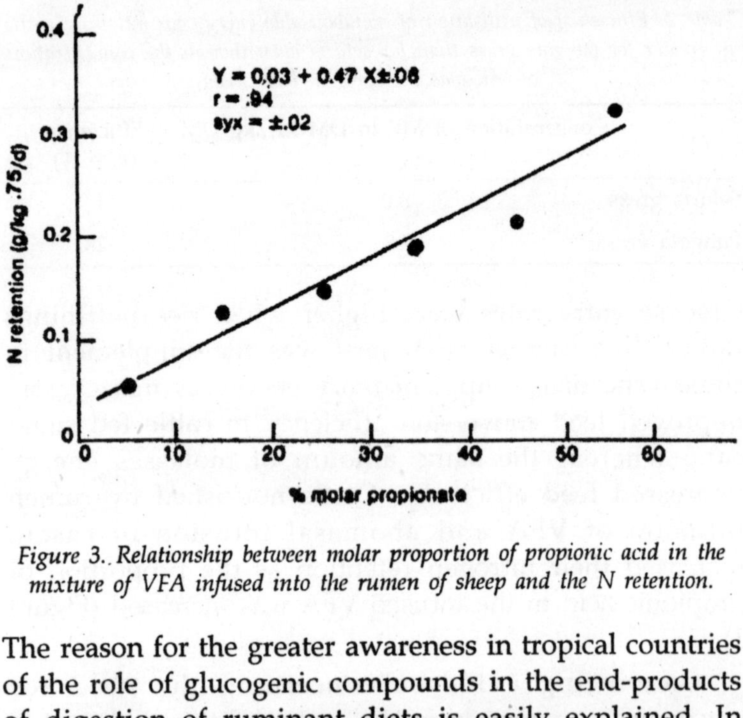


*Figure 3.. Relationship between molar proportion of propionic acid in the mixture of VFA infused into the rumen of sheep and the N retention.*

The reason for the greater awareness in tropical countries of the role of glucogenic compounds in the end-products of digestion of ruminant diets is easily explained. In almost all temperate countries, the diets for high producing ruminant animals contain at least 50 % of cereal grains. Such rations give rise to relatively large quantities of propionic acid in the rumen fermentation and there can be considerable escape of the starch for post ruminal digestion to glucose. Furthermore, most of the dietary nitrogen is in the form of true protein, and absorbed amino-acids can spare glucose.

There is thus rarely a deficiency of glucogenic compounds in the end-products of digestion. In contrast, tropical feed resources of high digestibility usually are of low glucogenic potential since they frequently contain carbohydrates (sugars) which are totally fermented, and give rise to VFA mixtures rich in butyric and acetic rather than propionic acid. Thus with feeds of temperate origin, increases in nutritive value are brought about by an increase in the proportion of cereal grain and oilseed meal in the diet and this gives rise to concommitant increases in the concentration of glucogenic compounds in the end-products of digestion. By contrast, in most tropical countries, cereal grains and protein-rich meals are in short supply and are competed for strongly for human and monogastric nutrition.

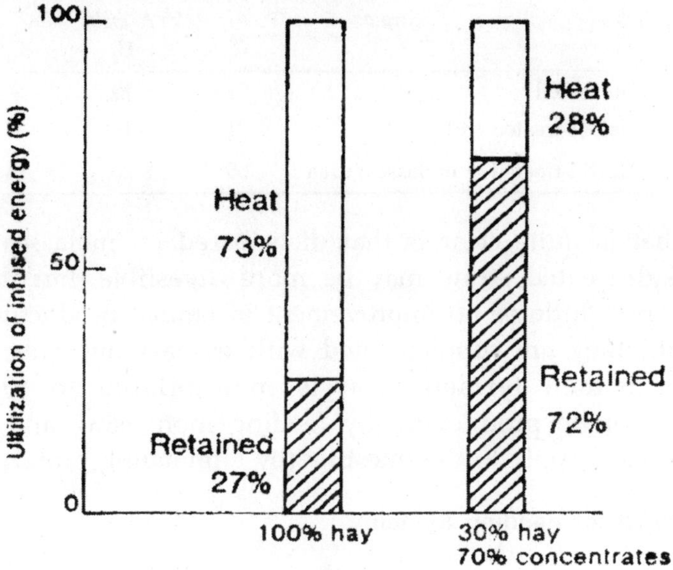


*Figure 4. Acetic acid (infused into the rumen) is utilised more efficiently when the diet has a high content of con- centrates.*

They are available for ruminant feeding in only small amounts. Improvements in digestibility of feeds in such regions are brought about either by the inclusion of sugar-rich by-products or by treatment with alkali/acid to hydrolyse resistant cell wall compounds. Raising the amount of sugar in the diet, or partially hydrolysing the cell walls in crop residues, does not lead to an increase in glucogenic compounds in the end-products of digestion, since neither of these strategies increases the liklihood of rumen escape of digestible carbohydrate. In addition, rumen fermentation of such modified feeds is characterised by relatively low proportions of propionate at the expense of butyrate or acetate Table 3.

*Table 3. Effect of alkali (NaOH) treatment of rice straw on VFA proportions in rumen fluid of cattle*

| *Straw* | *VFA molar %* | | |
|---|---|---|---|
| | $C_2$ | $C_3$ | $C_4$ |
| Untreated | 71 | 19 | 9 |
| NaOH-treated | 71 | 19 | 9 |
| NaOH-treated + molasses/urea | 65 | 20 | 14 |

What is quite clear is that diets based on molasses or alkali-treated straw may be more digestible, but they support little or no improvement in animal productivity until they are supplemented with by-pass nutrients; or the rumen fermentation is manipulated to raise propionate production (by feeding monensin), and/or protein supply to the intestine (by eliminating protozoa).

## Growth of Feeding Systems

### *Requirement of Nutrient for Production*

*The overall balance of nutrients*: The principal functions of large ruminants, where low quality pasture and crop residues are the main energy resource, are workpower, growth, pregnancy and milk production. Beef is very much a by-product being the residual or salvage value of the unproductive animals. In order to derive feeding systems for these different productive functions it must be appreciated that the nutrient requirements cannot simply be satisfied by optimising rumen microbial activity (i.e. total rumen digestion).

One of the most important objectives of ration formulation is to increase the voluntary intake of the basic resource, since this is frequently the first constraint. Secondly, as indicated earlier, it is the overall balance of nutrients available to the animal, not simply the products of rumen digestion, which determines productivity and feed utilisation efficiency. To understand this latter aspect, it is convenient to identify the specific nutrients that are needed for metabolism when the purposes of the feeding systems are to support such diverse functions as work, growth, pregnancy and milk production (Figure 5)

The greater part of the digested energy is undoubtedly needed for the animal to work both in the physical and metabolic sense. This can be defined as oxidation energy provided by nutrients which give rise to two-carbon fragments ($C_2$) catabolised in the Krebs or tricarboxylic acid (TCA) cycle. However, small amounts of energy in the form of glucose, or as compounds which can substitute for glucose and are glucogenic, are needed to catalyse the TCA cycle which is the mechanism by which oxidation energy is made available to the animal. There is an even larger requirement for these compounds to supply reduced cofactors (NADPH) needed for the elongation of the fatty acid chains in the synthesis of body fat and also milk fat.

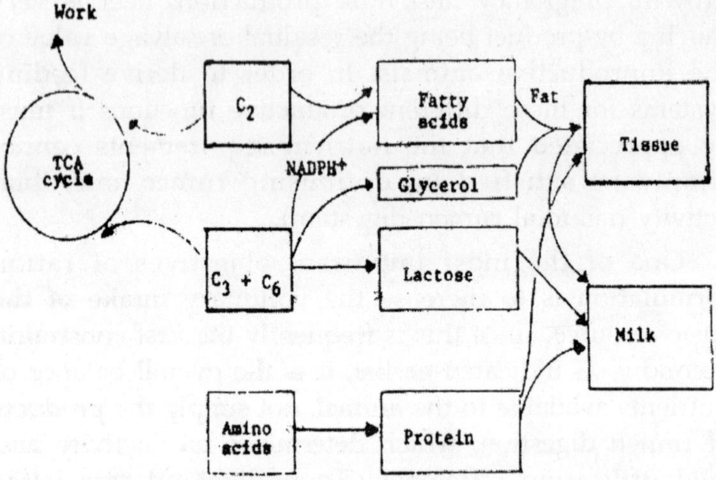


*Figure 5. Ruminant reaquirements for major metabolites according to productive state*

Glucose is the major energy nutrient for utilisation in the brain and the central nervous system. It is also needed for macro-molecule synthesis in tissue growth and to form glycerol in fats and lactose in milk. Glucogenic compounds may also spare the catalolism of some of the essential amino acids. For simplicity, glucose and glucogenic compounds can be referred to collectively as energy for synthesis or $C_3$ - $C_6$ energy; the remaining major nutrients are the amino acids which are also needed for synthetic purposes, namely of body tissues and the proteins in milk. All digested nutrients can give rise to oxidation energy but not all of them can supply glucogenic compounds and glucose or amino acids.

The main glucogenic precursor in ruminants is propionic acid produced in rumen fermentation. Glucose can also arise from post-ruminal digestion of starch; certain amino acids are also glucogenic and can provide precursors for glucose synthesis following deamination.

Amino acids are mainly derived from microbes grown in the rumen and the subsequent digestion of these organisms in the intestine; the second important source is from the post-ruminal digestion of dietary protein. These three groups of nutrients - oxidation energy, glucogenic compounds and essential amino acids - are required in different proportions according to the productive state of the animal.

The total requirement for energy is perhaps known with most precision and, for the purposes of ration formulation, can be described in terms of metabolisable energy (ME), or simply digestible energy (DE) intake, since ME is usually calculated from DE. The capacity of feed ingredients to give rise to glucogenic compounds, glucose and to the amino acids will depend on a number of factors many of which interact with each other (Figure 6). This makes it difficult to establish a simple additive system for ration formulation.

*Amino acids*: The supply of amino acids of microbial orgin to the animal will depend on the intake of fermentable feed and the efficiency of net microbial growth. The latter is controlled by a number of variables including rumen dilution rate, the presence of protozoa and the supply of precursors including fermentable N and minerals. For all practical purposes the amount of microbial protein derived from fermentable energy can be predicted on the basis of 3 g nitrogen per 100 g of digestible organic matter consumed.

The other source of amino acids to the animal which becomes increasingly important as the productive rate increases is the protein which escapes fermentation and contributes amino acids directly following digestion in the small intestine. The capacity of a protein meal to escape the fermentation is not only a function of the potential rate of breakdown of the protein by micro-organisms but it is also influenced by the dilution rate of rumen digestion as obviously the faster the flow of fluid and particles from the rumen, the greater the likelihood that they will escape rumen fermentation.

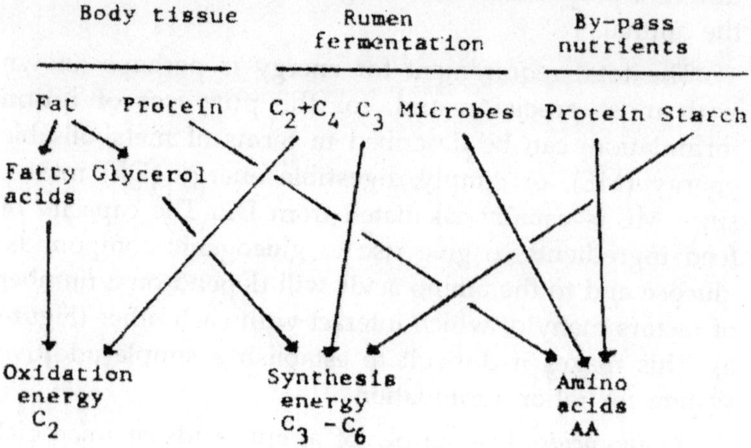


*Figure 6. Origin of principal metabolites in ruminants*

*Glucose and glucogenic compounds*: The dietary nutrients which give rise to glucose and glucogenic compounds are those that lead to high rates of production of propionic acid in the rumen fermentation. In diets containing starch, glucose may become available directly if the starch can escape to the lower tract. The availability of glucogenic compounds is also enhanced by the amount of protein reaching the intestine as a number of amino acids provide three carbon units capable of being converted to glucose.

Propionic acid production in rumen fermentation can be altered by chemical additives, such as monensin, which is widely used for this purpose in feedlot rations in Europe and North America. When amino acid requirements are high, glucose synthesis rates are high (Figures 7 and 8). The pattern of apparent requirements for glucose follows closely that for amino acids suggesting that part of the requirement for amino acids may be a requirement for glucogenic compounds. During growth and lactation, there may be competing needs for amino acids for glucose synthesis and for protein deposition.

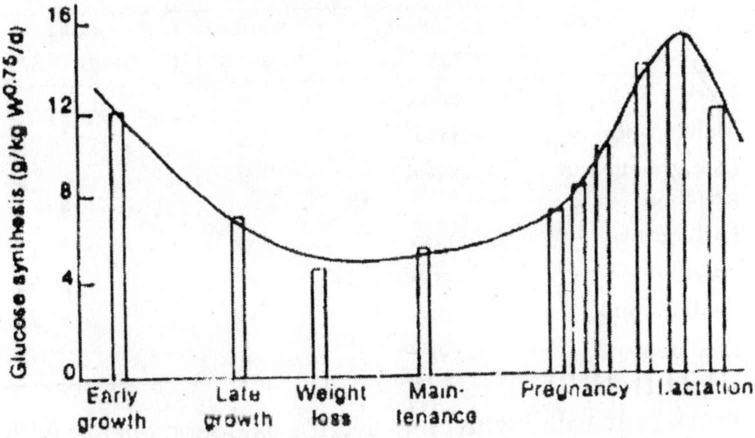


*Figure 7. Rate of glucose synthesis according to productive state in sheep*

The important point here is that in growing, pregnant or lactating ruminants there is a high demand for amino acids for protein deposition, and for both amino acids and propionate for glucose synthesis.

#### Requirements for Amino Acids and Glucogenic Compounds

There is insufficient information available to permit the precise quantification of the proportions of the different nutrients required at the metabolic level for different productive states. Nevertheless, an approximation of the needs can be attempted, based on a scheme attaching relative priorities to the groups of nutrients together with knowledge of the physiological and biochemical processes underlying the expression of the particular productive state. Such an indicative scheme is set out in Table 4.

*Table 4. Relative priorities attached to the principal metabolite groups according to the productive state of the animal*

| *Function* | *Relative priorities* | | |
|---|---|---|---|
| | *Oxidation energy ($C_2$)* | *Synthesis energy ($C_3$-$C_6$)* | *Amino acids (AA)* |
| Work | xxxxx | x | x |
| Maintenance | xxxx | x | x |
| Late growth and gestation | xxxx | x | xx |
| Early growth | xxxxx | xx | xxx |
| High milk | xxxxx | xxx | xxxx |
| Medium milk | xxxxx | xx | xxx |
| Low milk | xxxx | x | xx |

Work primarily involves a need for oxidation energy with minimum requirements for both glucogenic compounds and for amino acids. Maintenance alone obviously requires less energy expenditure. Late growth and gestation imply a relatively small increase in essential amino acid requirements. The requirements for growth in the young animal approaches that for high milk production, in terms of needs for amino acids and $C_3$ - $C_6$ relative to $C_2$ energy. Lower levels of milk production are obviously less demanding in terms of essential nutrients.

These priority ratings can be further simplified by assigning a ranking for each metabolite group according to the productive state being considered. Synthesis energy ($C_3$ - $C_6$) means the requirements for glucose and glucogenic compounds while the needs for amino acids are described in terms of supplementary by-pass protein (B-P) on the basis that the microbial protein supply will be determined by the energy status of the diet (Table 5)

*Table 5. Arbitrary rankings for synthesis energy ($C_3$-$C_6$) and by-pass protein (B-P) according to productive needs*

| Function | Rating (on scale 0 to 5) | |
|---|---|---|
| | $C_3$-$C_6$ | B-P |
| Work | 0.1 | 0.2 |
| Maintenance/gestation | 0.1 | 0.5 |
| Beef fattening | 0.5 | 1.5 |
| Low milk production | 1.0 | 1.0 |
| Medium milk production | 2.0 | 2.0 |
| High milk production | 3.0 | 3.0 |

### Nutritive Value of Feeds

The most important feed resources are:

- Dry season pastures
- Crop residues
- Low value agro-industrial by-products

These feed resources will usually form the basis of the diet and generally will not be limited in quantity. The need is mainly to identify an appropriate supplement which should be tailored according to animal needs for the particular function under consideration. One role of the supplement will be to correct/complement the balance of the digestion end products in terms of the supply of amino acids and glucogenic precursors. There is no adequate analytical means of assessing the potential value of feed ingredients as sources of amino acids, glucose and glucogenic precursors. The proportions of these nutrients in the digestion end products will be influenced by:

- the amounts of starch and protein in the feed
- the nature of the rumen microflora (ratio of bacteria to protozoa)
- the pattern of the fermentation (ratio of propionate to other VFA)
- the rate of fermentation of the feed
- the rate of passage of liquids and particles out of the rumen
- escape of starch and protein from the rumen.

The cheapest feed ingredients are the natural grazing and/or crop residues and, to a lesser extent, agro-industrial by-products which will constitute the energy sources; and non-protein nitrogen, in the form of urea and ammonia. The expensive feeds are the protein meals, derived from oilseed residues and the by-products from processing of animals, fish, and starch-containing cereal grains, which are also the staples of human and monogastric animal nutrition.

In general terms, therefore, sources of oxidative energy and fermentable nitrogen are relatively inexpensive, while the sources of amino acids and glucogenic compounds are very expensive. Since it is the combination of fermentable energy and fermentable nitrogen which gives rises to amino acids in the form of microbial protein, and as feeding level is associated positively with protein supply to the duodenum, it is generally desirable to supply the basal feed resource on an ad libitum basis.

The fermentable nitrogen requirement can be predicted on the basis that 100 g of fermentable carbohydrate in the feed will give rise to microbial protein containing 3 g of nitrogen. Of course, it is not always necessary to provide the total amount NPN, since a proportion of the feed protein will always be fermented to ammonia and in addition considerable blood urea-N may be secreted into the rumen. These processes reduce the amount of non-protein, fermentable nitrogen needed. The potential of the final mixed diet to satisfy the requirements of the animal for amino acids and glucogenic precursors depends principally on its content of protein-and starch-containing ingredients capable of escaping rumen fermentation and which are able to be digested in, and the products absorbed from, the intestine.

### Formulations of Feeding Systems

The capacity of feeds to provide oxidative energy can be estimated relatively precisely from the digestion coefficients. These can be determined by a number of means; by total collection of faeces from animals fed ad libitum; from the 24-hours DM loss from nylon bags in the rumen; or from some appropriate in vitro fermentation test. By contrast, it is almost impossible to predict accurately, from analysis of individual feeds, the effect that combinations of these feeds will have on the final supply of amino acids and/or glucogenic compounds, at the levels required for efficient metabolism in the animal. Two schemes are proposed in order to devise suitable supplementation programmes for achieving particular rates of productivity.

#### *Least-cost Complete Rations*

The first scheme is for the computerised formulation of least-cost complete rations. The available basic feed resources and supplements are rated according to their potential to give rise to amino acids and glucogenic compounds in the end-products of digestion (Table 6). This information is then combined with the nutrient requirements according to the productive state expected of the animal (Table 5). These two sets of data are then incorporated in a computer model for least-cost ration formulation together with the traditional standards for the anticipated level of production for digestible energy, calcium and phosphorus etc.

*Table 6. Rating of some feed ingredients according to their potential to supply by-pass protein and glucogenic compounds*

| | *By-pass protein (B-P)* | *Glucogenic compounds ($C_3$-$C_6$)* |
|---|---|---|
| Sugar cane bagasse | 0 | 0 |
| Straws (rice, wheat) | 0 | 0 |
| Elephant grass | 0 | 0 |
| Molasses (Sugar cane) | 0 | 0 |
| Sugar cane juice | 1 | 2 |
| Sorghum grain | 1 | 4 |
| Maize grain | 1 | 5 |
| Alfalfa/Berseem hay | 2 | 1 |
| Leucaena leaf | 2 | 1 |
| Wheat bran | 3 | 3 |
| Maize gluten meal | 4 | 4 |
| Soybean meal | 4 | 4 |
| Meat meal | 4 | 1 |
| Fish meal | 5 | 2 |
| Cottonseed meal | 5 | 4 |
| Rice polishings | 4 | 5 |

The limitations of this of this scheme are in the arbitrary assignments of the constraints for 3-$C_6$ energy and by-pass protein (B-P). The scoring is based on wide experience of, plus information from the literature on, the feeds in question. As such, the information is a form of substitute for "farming wisdom", which is probably the most important input in ruminant production systems in the industrialised countries.

A more empirical scheme, which is more appropriate for the conditions of most developing countries, is to select the basal energy resource according to physical availablility and price and then to provide supplementary nutrients in accordance with their relative priorities (Table 7) and costs. In this system, the first step is to provide a source of non-protein nitrogen to raise the level of fermentable nitrogen to a minimum of 3 % of the digestible organic matter.

It is desirable that this is done in a way which will ensure an almost continuous supply of ammonia-nitrogen to the rumen micro-organisms. The second priority is to give good quality green forage, preferably legume, up to a maximum of about 0.7 % (DM basis) of liveweight (about 25 % of the diet) daily.

*Table 7. Strategy for optimising ruminant productivity on carbohydrate-rich: low-protein diets*

A : Select basal feed resource

B : Supplement with:

(i) fermentable N (urea; $NH_3$)

(ii) high quality (leguminous) forage

(iii) by-pass protein

C : Increase degradability of cell wall component of basal diet (alkali/acid treatment)

D : Improve the glucogenic capacity of the diet by manipulating the rumen fermentation to produce more propionate; or by supplementing with by-pass starch.

Finally, a protein-rich by-product of oilseed crushing or cereal processing, or an animal by-product meal, should be given in amounts not to exceed 20 % of the total diet

DM. The 20% limit is to prevent depression/substitution of the digestible energy of the basal diet. Lesser amounts may be more economical, and it is imperative that response trials be carried out so that the amount of supplement can be related to the rate of animal productivity.

Two experiments carried out with agro-industrial residues composed almost entirely of soluble sugars and cell wall compounds and with extremely low contents of nitrogen and protein, corroborate the relevance of the above principle, and indicate that the best results from these feeds are achieved when all the supplements are combined into one feeding system. In the case of the derined sugarcane, animal response was increased from 0 to 500 g/day live weight gain by either high quality forage or cottonseed meal.

Nevertheless, performance was further increased by 100% to reach 1 kg/day when all the supplements were combined in the same ration. It is appropriate to note that high levels of urea only gave a response when both the good quality forage and/or by-pass nutrients were also provided. The second trial was with ensiled henequen (sisal) pulp. This feed resource is deficient in phosphorus and trace elements, highly imbalanced with respect to calcium and provokes a metabolic acidosis. There was no response in animal performance to supplements which neutralised the acidosis (Harrison D.G., personal communication) or which provided an appropriate balance of minerals (Figure 8).

However, limited amounts of either a good quality green forage (Brosimum alicastrum) or a by-pass supplement (soybean meal) changed the liveweight loss in lambs of -40 g/d to a gain of 55 g/d. When all supplements were combined, the growth rate was increased to 125 g/d.

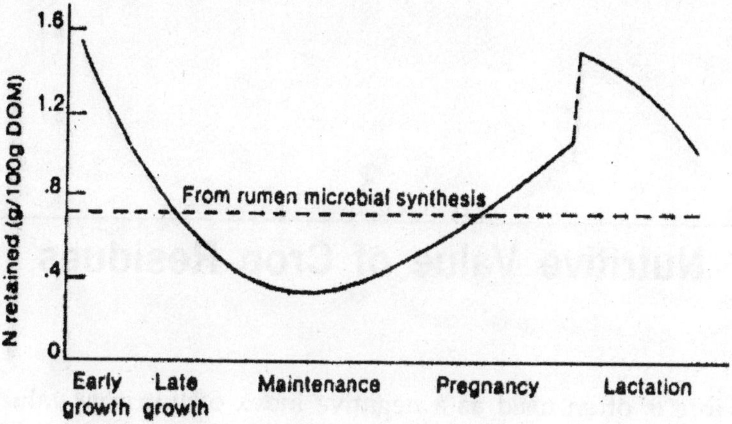


*Figure 8. Requirement for amino acids in sheep according to productive state*

As in the case of the urea in the previous trial, there was no response to minerals until both rumen function and by-pass nutrient status were provided for by the combination of green forage and soybean meal.

# 3

# Nutritive Value of Crop Residues

Fibre is often used as a negative index of nutritive value in the prediction of total digestible nutrients (TDN) and net energy. Prediction equations assume that higher fibre means lower digestibility. The association between fibre and digestibility is strong in temperate forages because of the strong association between lignin and cellulose in first cuttings in temperate climates. However, this association is weak in tropical forages, straws and stovers and fibre cannot be used to predict digestibility of these feeds.

The role of fibre analysis in the evaluation of feed quality is complex: the total amount of fibre (plant cell wall) is a negative index, while the amount of available fibrous carbohydrates is a positive index. Coarse fibre is required for normal rumen function and metabolism and is a positive dietary factor. The physical characteristics of fibre (particularly particle size) are also important in regulating rate of passage, rumination, insalivation and the pH of the rumen.

The total fibre or cell-wall fraction of plants comprises cellulose, hemicellulose, lignin, cutin, silica and a variety of minor substances. The proportions of these components vary among parts of the same plant and also change as plants mature.

Forage dry matter can be divided by means of the detergent system into a readily available soluble fraction and a fibrous residue of limited availability. The nutritive value of the fibrous fraction is determined by the degree of lignification, while that of the soluble non-fibrous portion is completely available to digestion. Utilisation of starches and other non-fibrous carbohydrates is limited only by the extent to which they escape the digestive process by passing rapidly through the digestive tract.

The value of chemical analysis for evaluating forages has been called into question on the basis that composition does not predict nutritive value, and that such analyses are too expensive relative to the amount of information provided. The critics would replace chemical analyses with nylon-bag degradability tests, and perhaps an analysis for nitrogen by Kjeldahl. They also charge that the originators of new chemical methods would convert all ruminant nutrition laboratories in the world to their procedures. This view ignores the purpose of detergent analyses which is to understand factors that limit the availability of energy and protein, and to correct the errors of proximate analysis and crude fibre by providing an accurate partitioning of cell-wall components into hemicellulose, cellulose and lignin.

Chemical analysis of feedstuffs is aimed at understanding why a given feedstuff exhibits its nutritional peculiarities. When this is understood, evaluation for causative and controlling factors can become routine, and not very expensive, since only the relevant factors are analysed. Unfortunately, the particular limiting nutritional factor varies among individual forages and feedstuffs, such that a single, universal analysis is not possible.

Most of the misunderstanding of the use of chemical analyses for estimating nutritive value has arisen from incorrect application of methods. Proper application of analyses is directed toward solving specific problems, and thus the set of particular analyses will vary with the experimental situation. The primary purpose of laboratory analysis is to characterise forages and feedstuffs so that nutritive value and performance in livestock can be related to chemical composition. This relationship has been the basis of compositional feeding tables for some time.

Compositional data should include those components most pertinent to nutritive value. These vary among feeds but, for many feeds, determination of cell wall, lignification and nitrogen or crude-protein content will be sufficient. Ash, lipid and available carbohydrate contents are the next most useful factors to determine. Proximate analyses including crude fibre and nitrogen-free extract (NFE) are inadequate for estimating nutritive value because they do not represent the components that limit biodegradability in the digestive tract.

Nylon bag degradability is also unsatisfactory, because it measures digestibility and is not a chemical entity. Nylon bag measurements reflect rumen environment involving diet and animal differences and are thus inherently more variable than in vitro measurements of digestion. The latter are recommended for plant breeding studies.

The detergent system of analysis is intended to provide a biologically realistic description of forages and fibre. The use of fibre values for predicting digestibility is called into question by the chemical and nutritional non-uniformity of these fractions and thus the problem of crude fibre and NFE is not only one of erroneous analytical fractionation, but also of application in the form of empirical regression equations that attempt to estimate digestibility from fibre content. The use of regressions of nutritive value on chemical composition assumes that nutritive value is determined by the chemical fraction measured.

Applied to fibre, the assumption is that fibre limits nutritive value. However, the digestibility of fibre is limited by lignification and fibre is only secondarily related to digestibility through its association with lignin. Since lignification is greatly affected by environmental conditions, such as temperature, daylength, light and plant stress, the association of fibre with lignin content is highly variable. This is reflected in a positive association between fibre and lignin in forages growing in the spring (when temperature and daylength are positively related) and a negative association in the autumn.

There is little or no association between fibre and lignin in forages growing in midsummer or under tropical conditions. Despite this, acid-detergent fibre (ADF) or modified ADF (MADF) are used as principal criteria for estimating digestibility in both America and Europe. The interactions of environment and climate with plant physiology and growth are sufficient to render associations between fibre components and nutritive value unreliable, and thus demolish the model that fibre regulates digestibility. The continued use of such regressions does not constitute a valid scientific operation and is an abuse of the respective methods of analysis.

Those who would justify empirical regression systems argue that ADF and NDF do not represent chemical entities and can therefore be treated like crude fibre. This overlooks the biological criteria for uniform chemical factors used in the establishment of detergent analyses, i.e. by the methods of Lucas. Crude fibre was intended as a determination of cellulose and, although imperfect by modern standards, is closely correlated with cellulose content.

Acid-detergent fibre is intended to represent lignocellulose and provide a preliminary step in the preparation of lignin, free from interference from protein. It contains small amounts of other cell wall components, viz bound protein and nitrogen, cutin, biogenic silica and micellular pentosans. However, these are of interest as unique fractions of very low digestibility. The determination of acid-detergent insoluble nitrogen (ADIN) for unavailable protein and the preparation of ADF for lignin and cellulose measurements are its main uses. Neutral-detergent fibre (NDF) includes hemicellulose and represents the insoluble plant cell wall matrix that is cross-linked with lignin.

The digestibility of the cell wall matrix depends on the extent of lignification. Pectin is not covalently bound to the cell wall matrix and is largely extracted by neutral detergent and is completely fermentable. NDF contains the indigestible lignified matrix and associated components of the cell wall and represents the skeletal structure and volume of the plant. It is the only plant fraction that can account for rumen fill and voluntary intake of forage and that is highly correlated with both rumination and chewing time among a wide range of forages.

## Tropical Forages, Straws and Stovers

The TDN of tropical forages averages 15 units lower than that of temperate forages due to effects of climate and management. However, reported maximum and minimum digestibilities from extreme immaturity to full maturity of forage grasses vary with latitude (Figure 1). The relationship between straw digestibility and latitude is probably similar to the minimum values in Figure 1.

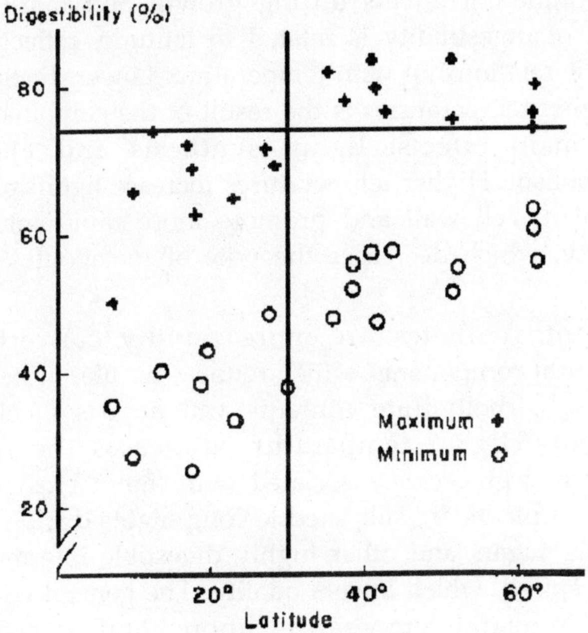


*Figure 1. The relationship between digestibilities of perennial grass and latitude.*

The tropics are the geographical regions that are free from frost and forages can grow continuously in this region if sufficient moisture is available. In temperate latitudes growth begins with the cessation of frost. Thus, maximum digestibilities are high and show no change with latitude above 30 degrees. However, in tropical areas growth begins at higher temperatures, usually after cutting or when rains end a dry spell. Survival of exposure to frost demands the accumulation of reserves to provide cold-tolerance, thereby increasing digestibility.

The same effects are seen at high altitudes in the tropics and in arid areas. The digestibility of plant material at maturity is determined by the cumulative environmental effects during growth and maturation. Level of digestibility is related to latitude, reflecting an inverse relationship with temperature. Lower digestibility at higher temperatures is the result of the combination of two main effects: lignin synthesis and elevated metabolism. Higher temperatures increase lignification of the plant cell wall and promote more rapid metabolic activity, which decreases the pool of metabolites in the cell.

Photosynthates are more rapidly converted to structural components, which reduces nitrate, protein and soluble carbohydrate contents and increases cell wall content. Higher temperature increases the rate of enzymatic processes associated with lignin biosynthesis. Tropical plants are subjected to long nights during which soluble sugars and other highly digestible intermediates are respired, which lowers quality. The general effects of environmental temperature upon plant growth and composition appear uniform in all plant species studied.

However, the quantitative effects of temperature upon forage quality vary with plant parts and with plant species. Temperature has its greatest overall effect on plant development in promoting the accumulation of lignified cell wall. This may be modified by growing conditions and species. For example, plant species that remain vegetative, whether by reason of too low environmental temperature during growth or by genetic character, are almost always less lignified that those plants that develop to the flower stage under similar environmental conditions (e.g. pangola grass).

A physiological reason is that lack of flowering and seed development allows the required resources to remain in the leaves and stems promoting higher nutritive value. This effect is very important in cereals grown under conditions of poor grain production leading to a straw of higher nutritive value. Another characteristic of tropical forages is the wide range in quality within the same standing plant.

Animals tend to select better quality material and the difference in composition between what is eaten and what is refused may be considerable. In straws and stovers the nutritional quality of the leaf may be considerably higher than that of the stem. Thus a major factor affecting quality of straw and stover is the recovery of leaves. An exception to this is the case of rice straw where leaves are of lower quality than stems. Straws and stovers are often chopped in order to reduce bulk, increase consumption and reduce wastage. This practice forces the animal to eat more of the low quality parts and reduces the nutritive value of that actually consumed.

More selective feeding on the part of the animal is a recognised feeding strategy of goats and sheep which are more limited by their smaller metabolic size. These animals can be adversely affected by chopping forages. High producing dairy cattle are also sensitive to the physical form of forage.

### Effects of Soil

Many cereals and grasses metabolise silica and deposit it in opaline form in the cell walls of leaves, reducing their digestibility. Thus availability of soil silica can affect straw quality. This effect is particularly striking in rice which contains up to 20% silica, which is selectively distributed in leaves and seed hulls. Rice straws are low in lignin and silica is the prime factor limiting digestibility.

### C3 and C4 Species

The first stable products of photosynthesis in many

tropical grasses are four carbon compounds, while those of dicots and most temperate grasses are three carbon compounds, hence the designation as C3 and C4 species. C4 plants are photosynthetically more efficient than C3 plants. Tropical C4 plants tend to accumulate large amounts of low-quality dry matter. These plants have fewer mesophyll cells between vascular bundles than C3 plants and, since mesophyll cells are comparatively unlignified and highly digestible, their proportion influences quality.

Not all C4 plants have lower nutritive value than C3 plants. Maize, for example, is a C4 plant developed from tropical ancestors. When grown in temperate regions its nutritive value is high. However, Deinum noted that maize grown under warm conditions reverts to many of its tropical ancestral characteristics and is of lower nutritive value (Table 1).

*Table 1. Effect of environmental temperature on the in vitro digestibility of mature leaves at final harvest in maize.*

| Temperature (°C) | 17/12 | 20/15 | 25/20 | 30/25 |
|---|---|---|---|---|
| OM digestibility (%) | 88.6 | 87.2 | 80.8 | 79.5 |
| Cell wall content (%) | 53.3 | 53.0 | 56.8 | 53.4 |
| Cell wall digestibility (%) | 80.1 | 77.8 | 69.4 | 66.4 |

C3 and C4 plants coexist in the tropics but C4 forage plants (mainly grasses) dominate favourable environments because they are more aggressive and higher yielding. The C3 tropical forages include the legumes and grasses adapted to less favoured conditions. As a group they are higher in nutritive value, but yield less and are less responsive to fertilisation. Agronomists in the tropics have favoured using the higher yielding C4 grasses, despite their lower quality.

### Fibre Quality

Fibre quality is defined as the ability to promote efficient rumen fermentation, and includes the potential digestibility and rate of fermentation of cellulosic carbohydrates, particle size and strength and cation exchange capacity. Graminaceous straws tend to be poor in these factors, although considerable variation exists. Legume fibre is superior in cation exchange capacity and rate of hydration to the average grass fibre, causing it to have shorter lag times after feeding and faster rates of fermentation. This occurs even in legumes of higher lignification.

The shorter lag and faster fermentation are associated with higher consumption. Grasses are characterised by high NDF and hemicellulose contents. As a result, intake of grass is lower than that of legume at a given digestibility. Grass fibre is also lower in lignin content than legume fibre. Thus, grass fibre is a better energy source for cellulolytic organisms than legume fibre, but its rate of fermentation and buffering capacity are lower. Lignin protects cellulosic carbohydrates from digestion but is responsible for much of the cation exchange capacity. Thus lignin may have both a positive and a negative effect on fibre quality.

### Forage Treatment

Treatment of low-quality forage to improve its nutritional value has usually employed alkali. Removal of the limitation of lignification on digestion depends on either cleavage of the bond between lignin and carbohydrate or hydrolysis of the polysaccharide away from the lignified matrix. Most studies of the effects of roughage treatment have measured animal responses but not chemical changes in the forage, and thus most procedural evaluations used at the present are quite empirical.

Analysis of lignin is the most obvious means of determining the efficiency of delignification.

However, the current practice not washing the straw (for economic reasons) allows the cleaved lignin to remain in the product. Also $NH_3$-treatment with heat can elevate apparent lignin content through polymerisation of carbohydrates and proteins in the Maillard reaction. Unfortunately, lignin analyses do not distinguish cleaved from uncleaved or synthetic lignin and neither lignin nor fibre content reflects the improvement in in vitro digestibility of treated straws.

Subsequent studies indicate that the relationship between lignin and digestibility is considerably different in treated straws. As a result, alkali- treated straws are often evaluated via rumen in vitro or cellulase digestion techniques. A chemical method to evaluate alkali-treated straw must distinguish cleaved lignin from uncleaved lignin. One procedure uses saponification of the isolated neutral-detergent fibre prepared without the use of sulphite. This measures the ester bonds left unhydrolysed after treatment with alkali.

The solution from neutral-detergent extraction can also be used to assess the quality of treated straws because cleaved lignin is soluble in neutral buffers and can be measured by UV absorption at 280 and 314 nm. Saponification with sodium hydroxide provides a solution which contains the residual lignin. Absorption at 280 nm results largely from phenolics, whereas ester linkages absorb at 315-340 nm; however, the latter wavelengths probably also include a component that is not ester related. A third possible analysis is to determine residual lignin on an acid-detergent residue that has been prepared by sequential extraction with neutral detergent followed by acid detergent.

The cleavage of lignin-carbohydrate ester bonds in graminaceous straws results in the release of phenolic compounds into the soluble fraction. The digestibility of the soluble fraction is depressed because the cleaved phenolics are not digestible. This may partly explain the difference between rumen *in vitro* and *in vivo* digestibility coefficients in treated straws. This difference has not been explained by any laboratory measurements, although it has been attributed to faster passage of the treated forages due to changed physical structure. McBurney examined 30 treated and 15 untreated samples of straws. Sixteen forages were treated with ammonia and 14 were treated with sodium hydroxide.

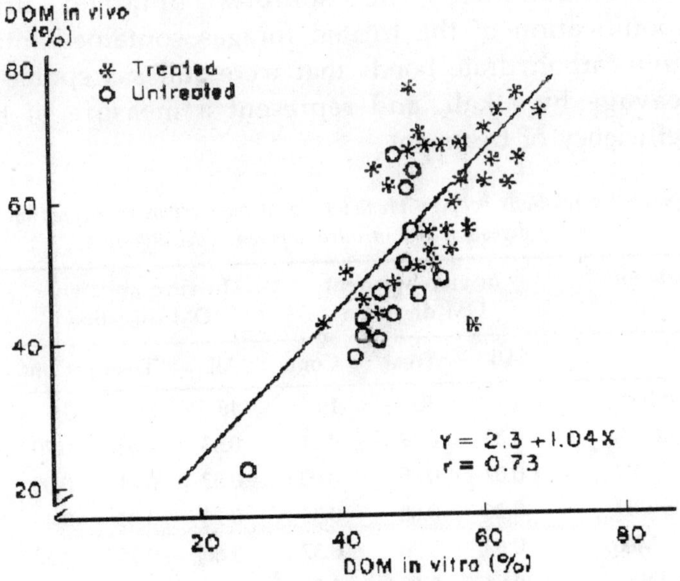


*Figure 2. The relationship between in vivo and in vitro digestibility of organic matter.*

Treatment tended to decrease NDF, but increased ADF, causing a drop in hemicellulose content. The effect was more pronounced with NaOH. The availability of additional nitrogen supplied by ammoniation varied widely: 0 to 66% of the additional nitrogen was in the acid-detergent indigestible nitrogen (ADIN) fraction which is unavailable to the animal and to the microbes. The relationship between *in vivo* and *in vitro* OM digestibilities is shown in Figure 2. *In vitro* measurements can both overestimate and underestimate *in vivo* values. The inverse correlation between lignin and digestibility was stronger in untreated than treated forages (Table 2).

The correlation between digestibility and optical density of neutral-detergent extracts, which measures cleaved lignin carbohydrate bonds, was positive in the treated samples. The solubles obtained after saponification of the treated forages contained intact lignin-carbohydrate bonds that were still susceptible to cleavage by alkali, and represent a measure of the inefficiency of treatment.

*Table 2. Correlations between predictors of in vivo apparent organic matter (OM) digestion and in vitro apparent OM digestion.*

| Predictor | In vivo apparent OM digestion | | | In vitro apparent OM digestion | | |
|---|---|---|---|---|---|---|
| | All | Treat. | Cont. | All | Treat. | Cont. |
| Number | 46 | 30 | 16 | 46 | 30 | 16 |
| Lignin % | 0.33 | 0.42 | 0.51 | 0.53 | 0.63 | 0.70 |
| NDS $OD_{280}$ | 0.51 | 0.35 | 0.00 | 0.52 | 0.41 | 0.09 |
| NDS $OD_{3142}$ | 0.26 | 0.08 | 0.35 | 0.23 | 0.08 | 0.46 |
| Sap. Meq. | 0.62 | 0.36 | 0.37 | 0.60 | 0.35 | 0.29 |
| SS3 $OD_{280}$ | 0.64 | 0.68 | 0.21 | 0.66 | 0.64 | 0.34 |
| SS $OD_{314}$ | 0.67 | 0.74 | 0.26 | 0.68 | 0.70 | 0.33 |

The improvement in digestibility due to treatment is highly variable. In some cases digestibility decreased, which appeared to result from moulding or other fermentation and an increase in net lignin and phenolic absorbance from the formation of Maillard products from heating. However, much of the variation in efficiency of treatment may be due to buffering capacity, which differed widely among the straws (Table 3). No current recommendation for straw treatment considers buffering capacity. Buffering capacity is related to cation exchange capacity, one of the criteria of fibre quality. Ironically, much of the exchange capacity is lost upon alkali treatment of fibre.

*Table 3. Buffering capacity means (X) and standard errors (SE) of a subset of control feeds.*

| Feed | n | X | SE | Coefficient of variation (%) |
|---|---|---|---|---|
| Cereal straw | 24 | 4.8 | 2.8 | 58 |
| Grass | 7 | 14.5 | 5.7 | 39 |
| Bagasse | 2 | 7.1 | 5.8 | 82 |

Generally, alkali treatments lack quality control and are expensive relative to the increase in nutritive value obtained. This severely limits their application, and it may be more realistic to supplement nutrient-deficient straws than to treat them.

## Buffering Capacity and Cation Exchange

Exchangeable groups in the plant cell wall include carboxyl, amino, nonhydrogen-bonded hydroxyls, and phenolic hydroxyls, all of which may bind metal ions. Thus the surface properties of fibre, i.e. hydration and cation exchange, are correlated (r about 0.7) and influence cell wall fermentation. The lag between ingestion and fermentation is related to how fast the fibre becomes hydrated and subsequent attachment of microbes. The amount of fibre digested after 6-12 hours of incubation is highly related to feed intake. Microbes have negatively-charged cell walls and recognise and attach to fibrous particles by their exchangeable surface.

This attachment requires formation of ligands between the microbial cell wall and fibre by divalent cations (probably magnesium). The cation exchange capacity of the fibre is its ability to bind and hold metal ions on its surface in much the same way that clay minerals hold cations in soil. The exchange serves as a bank exchanging K, Ca, Na, Mg for hydrogen when the pH drops and recharging as cations become available when saliva and ingesta are mixed. An advantage of this regenerable bank is that ruminated fibre passing down the digestive tract contributes buffering action further down the gut. The buffering capacity of feedstuffs derives in part from the physical effects they elicit in the rumen and during rumination and ensalivation.

Eating and rumination promote salivary flow containing much buffer that neutralises acids produced in fermentation. Fermentation of carbohydrates results in production of large amounts of organic acids, which must be removed by absorption to maintain pH and the normal rumen environment. Recycling of mineral ions is also important in maintaining rumen pH. The more slowly digested solid matter-fibre-contributes most to the maintenance of normal rumen environment.

Tropical grasses and straws have low exchange and buffering capacities, and supplementation with starchy concentrates renders the rumen sensitive to acidotic conditions, which reduce rumen efficiency and net feed intake. Tropical legumes and citrus are high in exchange and buffering capacity and are useful supplements.

Quality forage fibre is also associated with microbial efficiency. Fibre-digesting bacteria manufacture more cellular protein than do starch-digesting bacteria because of their lower maintenance costs, the relatively high ATP yield of acetate fermentation and evolutionary selection for cellular storage. Rumen organisms have substrate preferences and can divided into competitive consortia of mutually symbiotic species. For the purposes of this discussion they can be conveniently divided into:

(1) fibre digesters and

(2) those that specialise in starch (such as *Streptococcus bovis*) and more soluble carbohydrates.

The second group tends to be adventitious, producing lactic acid at the expense of cellular efficiency. Rapid production of lactic acid reduces pH and thus renders the environment more favourable for their growth because low pH is more inhibitory to the slower digesting organisms dependent on cellulose and hemicellulose. The rate of carbohydrate digestion is set by the physicochemical limitations of the substrate, which in turn limit bacterial efficiency. The combination of slow fermentation rate and competitive passage leads to inefficient use of the potentially digestible carbohydrates, with low microbial output and increased faecal losses.

There are important relationships between fermentation rates of carbohydrates and microbial efficiencies, i.e. production of microbial protein per unit of feed digested in the rumen. The fermentation rate sets the amount of feed energy available to rumen bacteria per unit time. Faster digestion provides more food, which dilutes the energy costs of maintenance and leaves more energy for growth and production. Bacteria have a maintenance requirement that must be met before growth can occur.

The maintenance requirement of cellulolytic bacteria is one sixth that of bacteria that ferment soluble starch and sugar. The type and quality of carbohydrate can have considerable impact on rumen microbial yield because carbohydrates are degraded at different rates. Cellulolytic bacteria grow slowly and a decrease in fibre quality can dramatically reduce yield by reducing their rate of digestion and growth. This limits the utilisation of straw, stovers and tropical grasses, which tend to be slowly digested.

## Phenolics in Fibrous Crop Residues

Phenolics occur in several forms in plants:

(1) as soluble compounds extractable with water (garlic acid esters) or with methanol and aqueous acetone (proanthocyanidins, flavonols, flavonolglycosides) and

(2) in non-extractable forms.

Phenolics may remain in the residue after extraction as a result of their inherent insolubility, e.g. due to their large molecular weights, or because they are covalently bonded to, or tightly bound in complexes with, other plant constituents.

### *Types of Phenolic Compounds*

Several types of phenolic compound affect the digestion of crop residues, including simple phenolic compounds, such as the cinnamic acids (I) or aldehydes (II) and polyphenolics, such as the 'condensed' (III) and 'hydrolysable' tannins (IV) (Figure 3).

#### *Polyphenolics*

"Vegetable tannins" are a group of polyphenolics, consisting of a large number of structurally very different compounds. The original definition of a "tannin" is a compound "able to convert hide to leather. "All tannins contain phenolic groups, but not all phenolics are tannins. Many of these compounds have now been identified. It is increasingly obvious that even the modified definition of a tannin as "a compound that binds to proteins" is imprecise and misleading and no longer serves any useful purpose. We will, therefore, try to avoid the word "tannin," except when referring to "condensed" and "hydrolysable tannins."

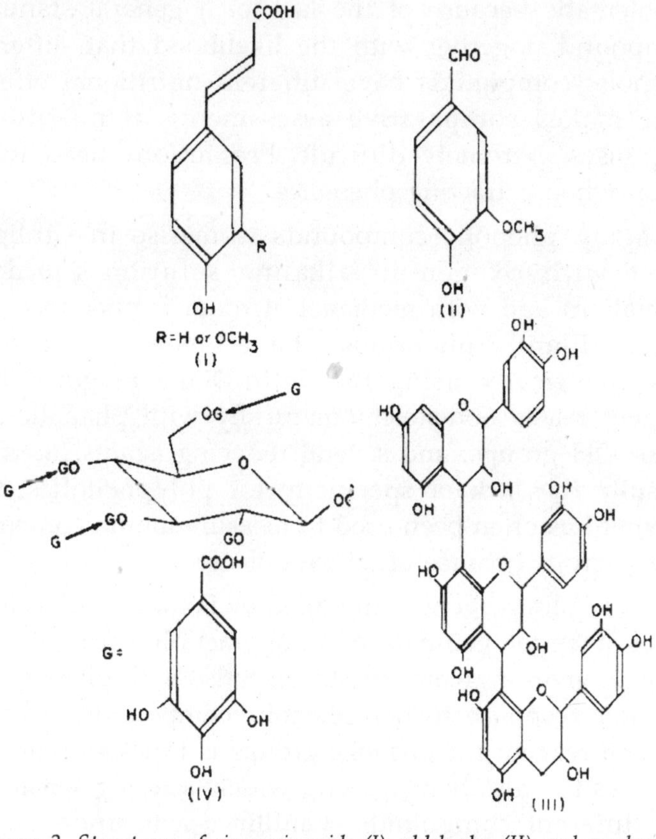


*Figure 3. Structures of cinnamic acids (I), aldehydes (II) and condensed (III) and hydrolysable tannins(IV).*

The literature on the effects of phenolics in animal nutrition has many examples of inappropriate use of analytical methods resulting in unsupported conclusions. This often stems from a poor understanding of the reactions between phenolic compounds and their detection agent. However, it may also frequently be a result of the use of unsuitable preparation procedures.

Different types of phenolic have been estimated using unspecific methods. Quantification of phenolics is problematic because of the lack of a general standard compound, together with the likelihood that different phenolic compounds have different nutritional effects. This makes comparative assessments of nutritional responses extremely difficult. Precautions need to be taken when extracting phenolics.

Many phenolic compounds isomerise in sunlight, react with oxygen in alkaline solution (quinone formation) and with methanol at room temperature and pH6. Plant phenolics have been measured colorimetrically using the Folin-Denis reagent. This reagent reacts non-stoichiometrically with phenolic and other OH groups, and several reducing agents interfere. Despite the lack of specificity for polyphenolics, this reagent has often been used to measure tannin content in forage crops because of its ease of use.

The Folin reagent as modified by Folin and Ciocalteau gives a better estimate of total phenolic groups. This reagent gives a greater colour response with phenols and a lesser response to non-phenolic compounds. Another general reagent for phenolic groups is Prussian blue, but this has the disadvantage of a widely varying sensitivity for different compounds. Vanillin reacts under acidic conditions with one group of phenols, called flavanols.

When used under the conditions recommended it is specific for a narrow range of flavanols (including condensed tannins) and dihydrochalcones. Heating proanthocyanidins with hydrochloric acid and n- butanol produces strongly coloured anthocyanidins. However, chlorophylls interfere and the reaction is not quantitative. The yield and type of anthocyanidins differ for each type of proanthocyanidin. Its advantage is its specificity.

In order to assess the effects of phenolics in nutritional studies, quantitative estimates of the amounts present are required. Gallic acid, tannic acid, catechin and chlorogenic acid have all been used as reference standards. However, tannic acid is a poor standard because it tends to be a mixture of different compounds, the relative proportions of which vary between samples. Using catechin as a standard in the vanillin test over-estimates the concentration of condensed tannins because the reaction kinetics are different for the monomer and the oligomers.

Gravimetric procedures have been used to obtain a direct measure of phenolics thus avoiding the problems associated with quantification in colorimetric assays. The hide powder assay measures the amount of phenolics adsorbed by the powder. However, Laurent found that results were variable and Verzele et al criticised the choice of hide powder because it selectively binds only some compounds from tannic acid mixtures. Reed et al developed a method for precipitating soluble phenolics from plant extracts with ytterbium acetate.

Precipitation was complete when phenolics accounted for more than 16% of the dry matter. The precipitates contained flavan-3-ols, condensed tannins, garlic acid and hydrolysable tannins, catechin gallates, flavonols and their glycosides. Little protein or chlorophyll appeared to be co-precipitated with the phenolics. This method holds promise for the gravimetric determination of phenolic compounds and modifications are being made to obtain complete precipitation of all phenolics at lower concentrations. High-performance liquid chromatography (HPLC) is better able to distinguish among different types of phenolics but separations may be incomplete with complex mixtures of similar compounds, e.g. some mixtures of condensed and hydrolysable tannins.

The metabolic role of phenolics in the plant is still not clear. Feeny suggested that they serve as chemical defence agents through their astringent taste and by interfering with the digestive enzymes of the predators. Although a negative correlation was found between susceptibility to bird attack and polyphenolic content of sorghum grain, Coley found no correlation between the extent of herbivore attack and phenolic content of mature leaves in a tropical forest. Similarly, Bernays stressed the variability of effects of polyphenolics on insect herbivores and warned against generalisation concerning their ecological significance.

A systematic approach by Beart et al noted that the principal biosynthetic thrust in plants is towards the production of higher molecular weight polyphenolics which are the end-product of the metabolism. Based on the "defence hypothesis" one might assume that such end-product are produced by plants as these compounds are most effective in precipitating the proteins of the herbivores. Beart et al therefore investigated the binding strength between various phenolic precursors and end-products with a protein, bovine serum albumin (BSA). To their surprise, they found that one common precursor, pentagalloyl glucose, bound the protein much more strongly than most of the end-products. This led them to doubt the defence hypothesis.

In a similar context in micro-organisms, Bu'Lock suggested that the secondary metabolism serves to maintain basic metabolism in circumstances not propitious for growth (e.g. nutritional imbalances). This is supported by the observation that the condensed tannin content of Lotus species was higher under low soil fertility than high soil fertility conditions and that high phenolic contents have been linked with high light intensity, high temperatures and severe drought. A further example is that sainfoin grown in the UK had a much lower extractable phenolic content than sainfoin grown in Ethiopia. The basic principles of the metabolism of phenolics must be better understood in order to develop strategies for the development of plants with higher nutritive value. For example, if stress conditions cause more phenolic precursors to be synthesised and specific plant enzymes cause their polymerisation, two screening approaches could be employed:

1. To assess the effect of stress on precursor synthesis in different varieties; and
2. To assess the differences in plant enzymes which produce the polyphenolics.

This approach is supported by the fact that different types of precursor occur in different sorghum plants, and by the discovery of different end products within the Cinnamon species. It should be possible to exploit such genetic differences in breeding programmes once the nutritional effects of the various phenolics are better understood.

### *Carbohydrates and simple phenols*

Several simple phenolics occur in plants. The derivatives of cinnamic acid are the most abundant; derivatives of benzoic acid and aldehydes occur in smaller amounts. Ferulic and p-coumaric acids are esterified to carbohydrates in plant cell walls, whereas the aldehydes are apparently linked at their phenolic groups to cell-wall polysaccharides. It has been suggested that phenolic compounds limit the digestion of carbohydrates: digestibility of cell-wall carbohydrates is increased when phenolic compounds are released from graminaceous forages by treatment with alkali. Very little is known about the type of bonding between polyphenolics and carbohydrates.

Beart et al proposed that some proanthocyanidins may be covalently linked via ether bridges at C-4 to carbohydrates, in analogy to the C-C bridges in proanthocyanidins, but the evidence for such linkages is circumstantial. Other types of bonding between polyphenolics and carbohydrates have been demonstrated; $H^+$-bridges and hydrophobic interaction binding are important in such complexes. Ford observed low digestibilities of cell walls from Desmodium intortum and suggested that this was probably caused by proanthocyanidins complexing with cellulose.

McManus *et al* studied binding among several polysaccharides and polyphenolics and concluded that the molecular size of the polyphenol and its conformational flexibility are important to the binding, which seems to be pH independent. They also noticed that small changes in the structure of either the polyphenol or the polysaccharide resulted in marked changes in their affinity for each other. Where the association was primarily a surface effect, broad similarities were noted with the analogous complexation with proteins. However, where the polysaccharide was also capable of forming inclusion complexes, significant differences were observed relative to the binding of proteins.

*Proteins*

The interaction between polyphenolics and proteins has been recognised for much longer than that with polysaccharides. However, we still do not understand the causes of the differences in the effects of polyphenolics among digestion trials. Polyphenolics have been reported to have both beneficial effects and detrimental effects on protein metabolism in ruminants. The complexes formed by the interaction of proteins and phenolics in solution may be either soluble or insoluble. However, insoluble (non-extractable) phenolics can also complex proteins. The effect of complex formation on the digestibility of proteins/phenolics and the activity of enzymes is not fully understood.

Mole and Waterman reported that condensed tannins both stimulated and inhibited digestion of complexed proteins by trypsin. Although several enzymes are inhibited by polyphenolics, others retain some or most of their activity whilst complexed. Thus, enzyme inhibition is not a good measure for so-called "tannins." The formation of complexes depends on the concentration of both the polyphenolic and the protein, resulting in variable stoichiometries.

Precipitation is thought to occur when a hydrophobic outer layer is formed. Thus at appropriate concentrations even simple phenolics, such as pyrogallol and resorcinol, can precipitate proteins from solution. This example illustrates best why the term "tannin" is obsolete. pH also affects precipitation. Precipitation is greatest at a pH within one unit of the isoelectric point of the protein. Martin et al demonstrated that metal ions influence the extent of precipitation between hydrolysable tannins and leaf fraction I protein: $Mg^{2+}$ and $Ca^{2+}$ were more effective than $Na^+$ and $K^+$ at bringing about protein precipitation. Our present understanding of the mechanism of interaction favours both $H^+$ -bond formation and hydrophobic interactions.

As indicated earlier for carbohydrates, complex formation is dependent on both solution conditions and the properties of the phenolics and proteins. Molecular size and conformational flexibility have major effects on the strength of binding between polyphenol and protein. For example, molecular flexibility of certain ellagitannins is less than that of gallotannins due to intramolecular crosslinking: as a result, the ellagitannins bind BSA more weakly than do gallotannins. Proteins with open, loose conformations interact much more strongly with sorghum polyphenolics. Thus, the great specificity of some polyphenolics for certain proteins can be likened to the specificity between enzymes and substrates. It follows that some proteins can be preferentially precipitated out of solution even in the presence of excess amounts of other proteins.

It seems reasonable to assume that the strength of binding in such complexes will have important implications on the degradability of complexes at different pH values, by enzymes or micro-organisms. Studies are also needed to assess any reactions within the polyphenol-protein complex that may occur during digestion. Beart *et al* predicted that covalent bonds could be formed in such complexes in the ruminant.

### Effect of Polyphenolic on Voluntary Feed Intake

Intake of feed containing large amounts of condensed tannins it low. Barry and Duncan recorded an increase in both metabolisable energy intake (MEI) and digestible organic matter intake (DOMI) in sheep fed high-polyphenolic Lotus in response to decreasing condensed tannin content when polyethylene glycol (PEG) was used to bind the condensed tannins. Digestibilities of OM, cellulose, hemicellulose and nitrogen also increased. Thirty-two percent of the increase in DOMI could be attributed to increased intake of digestible fibre and a further 32% to increased intake of digestible crude protein. However, an increase in intake of digestible crude protein does not necessarily increase supply of amino acids to the animal, as shown by Thomson et al working with dried lucerne and sainfoin fed to sheep.

#### Effect of Polyphenolics on the Sites of Carbohydrate and Protein Digestion

The presence of condensed phenolics in sainfoin and Lotus species has been associated with increased nitrogen retention in sheep. This has been attributed to an increased supply of amino acids to the small intestines as a result of protection of the plant protein from proteolysis in the rumen. In Lotus species condensed phenolic contents up to 25 g $kg^{-1}$ DM appear to have little effect on rumen carbohydrate digestion but concentrations between 25 and 100 g kg- DM reduce carbohydrate digestion in the rumen in a dose-dependent manner.

Barry and Manley found that, in Lotus based feed, increased polyphenolic content significantly reduced the extent of digestion of OM in the rumen The digestive behaviour of the high-polyphenolic Lotus was compared with the predicted behaviour of the same crop assuming it was a low-polyphenolic crop. The predicted increased flow of organic matter (measured as the sum of its individual components) was 99 $gd^{-1}$, of which 42% was fibre and the remainder largely crude protein. This supported the suggestion by Barry and Duncan that polyphenolics both reduce digestion of carbohydrates in the rumen and increase protein outflow. The amount of fibre excreted in the faeces was equal to the amount of extra fibre entering the small intestine suggesting that dietary fibre not digested in the rumen was irreversibly bound by polyphenolics.

In contrast, approximately 60% of the extra protein entering the duodenum was digested in the small or large intestine. However, in further similar experiments Barry et al found that the reduction in carbohydrate digestion in the rumen was compensated by increased post-ruminal digestion and whole tract digestibility appeared unaffected. In another study, Thomson et al examined the digestion of dried sainfoin (high condensed tannin) and lucerne (low condensed tannin) in mature sheep. Although the crops were harvested at similar stages of growth and had similar total N contents, cellulose digestibility was significantly higher on sainfoin (78 vs 67) and the extent of cellulose digestion in the rumen was unaffected. In contrast, apparent N digestibility was significantly lower in sainfoin (68 vs 51).

However, duodenal amino acid supply from sainfoin was 49% greater (178 vs 119 g $d^{-1}$) and, despite a small increase in total amino acid flow at the ileum (sainfoin 64, lucerne 52 g $d^{-1}$), availability of amino acids in the small intestine was 46 $gd^{-1}$ greater from sainfoin than from lucerne. This response was seen despite a significant reduction in apparent N digestibility (whole tract) and indicates (a) the futility of using apparent N digestibility as a measure of protein value, and (b) a marked positive effect of condensed tannins on protein supply to the host animal.

In this study, no real attempts were made to determine the origin of duodenal protein on the two diets, but in a parallel publication Harrison *et al*, using DAPA, reported that bacterial N contributed 37% (sainfoin) and 79% (lucerne) to total duodenal N, suggesting that undegraded dietary protein made a major contribution to the overall increased protein flow on sainfoin. Irrespective of its origins, the apparent availability of amino acids in the small intestine appeared to be unaffected by the presence or absence of tannins.

In a subsequent study, Beever and Siddons examined the effect of sainfoin tannins on 'protein protection' in the rumen, and the possible effect of a small proportion of sainfoin on the digestion of a tannin-free legume such as red clover. All diets had similar total N contents, but soluble N contents were noticeably lower on the sainfoin-rich diets (Table 4). Duodenal amino acid flows were less than amino acid intake (as expected on fresh forages), but were 11-28% higher on the sainfoin diet than on all other diets (Table 5).

*Table 4. Nitrogenous fractions of diets containing varying proportions of sainfoin and red clover.*

| | Percentage contribution (DM basis) | | | |
|---|---|---|---|---|
| Sainfoin | 100 | 40 | 20 | 0 |
| Red clover | 0 | 60 | 80 | 100 |
| | Dietary characteristics (g $kg^{-1}$ DM) | | | |
| Total N | 34 | 37 | 38 | 38 |
| Rumen liquor | | | | |
| TCA soluble N | 12 | 18 | 21 | 22 |
| Buffer soluble N | 8 | 11 | 14 | 15 |

Similarly, the flow of microbial protein was highest on the sainfoin diet and it was concluded that much of the increased duodenal. amino acid was due to increased microbial synthesis. This contrasts with the findings of Harrison et al, but in their study the diets were artificially dehydrated prior to feeding. Of possibly greater consequence, is that there was no positive interaction between sainfoin and red clover, from which it may be concluded that the polyphenolics of saifoin are specific (both physically and chemically) to sainfoin protein or that they were not present in sufficient quantity to exert any real effect on the red clover proteins.

*Table 5. Duodenal digesta amino acids flows and efficiencies of microbial synthesis in sheep receiving diets containing varying proportions of sainfoin and red clover.*

| Percentage contribution (DM basis) | | | | |
|---|---|---|---|---|
| Sainfoin | 100 | 40 | 20 | 0 |
| Red clover | 0 | 60 | 80 | 100 |
| **Duodenal amino acid flow (g $g^{-1}$ intake)** | | | | |
| Total | 0.83 | 0.65 | 0.75 | 0.67 |
| Microbial | 0.41 | 0.33 | 0.35 | 0.32 |
| Feed | 0.32 | 0.25 | 0.32 | 0.28 |
| Microbial synthesis (g $g^{-1}$ degraded amino acid) | 0.61 | 0.44 | 0.51 | 0.44 |

The effects of condensed tannins on carbohydrate digestion have been attributed, at least in part, to complexing of microbial extracellular enzymes. Condensed tannins inhibit in vitro proteolytic, cellulolytic and general fermentative activities of rumen microbes and in vivo microbial multiplication. No information appears to be available on the effects of condensed tannins on protozoal or fungal fermentative capacities, both of which may contribute considerably to digestion of structural carbohydrates.

However, the reduction in rumen digestibility in animals receiving diets high in condensed tannins may simply reflect insufficient rumen-degradable nitrogen (RDN) for maximum microbial activity. This is supported to some extent by the responses of rumen digestion to additional non-protein nitrogen. Adding 3% urea to a diet of wheat straw and deoiled sal seed meal fed to calves increased total nutrient digestibility from 48 to 61%. Similar effects were observed in vitro where adding urea to high-and low-condensed-tannin sorghum grains increased in vitro dry matter digestibility (IVDMD) from 73 to 93% in low-condensed-tannin cultivars and from 46 to 79% in high-condensed-tannin cultivars.

It is not known whether urea destabilises the bonding between protein and condensed tannins, releasing protein for microbial use, or if the urea simply acts as a source of RDN. The latter would seem to be more probable: Schaffert et al observed no effect of urea on in vitro protein degradabilities of sorghum grains despite increased IVDMD. In vivo studies also tend to support the idea that lack of RDN reduces digestion. Egan and Ulyatt associated the higher N retention in sheep receiving sainfoin than in those receiving clover or ryegrass diets with an increased rate of recycling of nitrogen (presumably as urea) into the rumen.

Barry et al reported rumen ammonia levels of about 16 mM in sheep receiving Lotus diets containing 95 g condensed tannins per kg DM and Reed and Soller found that sheep receiving acacia browse, which is high in condensed tannins, excreted significantly less nitrogen in the urine than sheep receiving Sesbania sesban or vetch hay, which are low in condensed tannins. The latter authors suggested that changes in urinary nitrogen excretion could be the result of increased microbial utilisation of endogenous nitrogen in the rumen. Digestibility of structural carbohydrates in the rumen is greater with protein supplementation than with NPN (urea) supplementation and responses to different protein sources differ. However, supplementing diets containing large amounts of condensed tannins with protein may not be economical and other ways to reduce the adverse effects of phenolics must be explored.

Other studies on nitrogen utilisation of feeds high in phenolic contents have shown both positive and negative effects, indicating the need for caution in extrapolating reported effects of phenolics on digestion or utilisation parameters from one plant to another. Reed et al examined the effects of phenolic compounds from a variety of plants in an *in vitro* cellulase system a and found that the degree of inhibition varied from about 2 to 70%: this variation was probably related to the types and amounts of phenolic compounds present. Other workers have also reported markedly different responses in rumen microbial activity to phenolics from different sources.

### *Effects of Plant Phenolics on Rumen Microbial Activity*

Recent experiments assessed the anti-microbial and physiological effects of several types of polyphenolic compound from Ethiopian browse plants on *Streptococcus bovis*. In general, all extracts increased lag times and reduced biomass yields. Growth rates were also reduced, but only in cultures grown with insufficient organic nitrogen. Large differences were observed in the extent of these anti-microbial effects when comparable concentrations of phenolics from *Acacia nilotica, Euclea schimperi* and *Pterolobium stellatum* were added to *S. bovis* cultures.

Extracts from *A. nilotica* were substantially more toxic, confirming that polyphenolics are not a uniform group of chemicals having similar effects. The effect of polyphenolics on *S. bovis* was reduced by increasing the organic nitrogen concentration in the culture medium (Figure 4). Other workers have also reported markedly different responses in rumen microbial activity to polyphenolics from different plant sources. The toxicity of free phenolic acids has been demonstrated in a wide range of rumen bacteria.

Bacteriocidal and bacteriostatic effects have been demonstrated in cellulolytic bacteria, such as *Ruminococcus albus, R. flavifaciens* and *Bacteroides succinogenes,* in the presence of 10 mM or less of p-coumaric acid and ferulic acid. Akin also observed that cellulolytic and xylanolytic bacteria were inhibited by p-coumaric acid and ferulic acid and that p-coumaric acid reduced motility in entodinomorph, but not holotrich, protozoa. Ferulic acid and sinapic acid had lesser effects on protozoa.

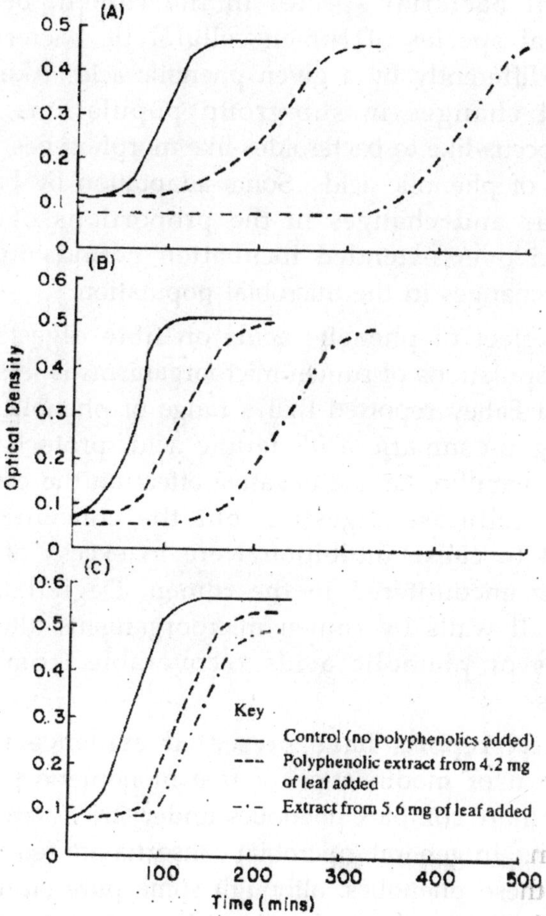


*Figure 4: Effect of polyphenolics from Acacia nilotica on the growth of Streptococcus bovis at different concentrations of organic nitrogen. Cultures were grown on glucose in medium B. A: no organic nitrogen (yeast extract and trypticase omitted). B: half the amount of yeast extract and trypticase C: full amount of yeast extract and trypticase.*

Vanillin and, to a lesser extent, cinnamic acid also inhibit cellulolysis by rumen bacteria, apparently by preventing them from attaching to cellulose particles. Similar observations have been reported by Akin *et al.* Differences in responses to phenolic acids may reflect the dominant bacterial species in the rumen, because individual species of rumen cellulolytic bacteria are affected differently by a given phenolic acid. Akin *et al* reported changes in sub-group populations, from ruminococcus-like to bacteroides-like morphotypes, in the presence of phenolic acids. Some adaptation by bacteria can occur and changes in the proportions of VFAs produced over extended incubation periods indicate possible changes in the microbial population.

The effect of phenolic acids on fibre digestion by mixed populations of rumen microorganisms is less clear. Jung and Fahey reported that a range of phenolic acids, including p-coumaric acid, ferulic acid, protochatecuic acid and vanillin, have a negative effect on the extent of in vitro cellulose digestion but the concentrations required to cause inhibition were in excess of those normally encountered in the rumen. Degradation of forage cell walls by rumen microorganisms alters the amounts of phenolic acids recoverable from plant materials.

Several reports have presented evidence for the breakdown or modification of free monomeric phenolic acids or more complex phenolics under strictly anaerobic conditions. In general, microbial consortia are required to modify these phenolics, although some pure cultures of rumen bacteria degrade phenolics, with a greater apparent degradation of ferulic than coumaric acid. Low recoveries of phenolic acids added to *in vitro* incubations have also been observed by a number of workers. Since cellulolytic bacteria are closely associated with structural polysaccharides in the rumen they may encounter locally high concentrations of potentially toxic bound and free phenolic acids as degradation proceeds.

No obvious relationship was found between depressions of cellulose and hemicellulose digestibilities and changes in substitutions to the aromatic ring of the phenolics although aldehydes were found to be more inhibitory than related acids. Hemicellulolytic bacteria appear to be more tolerant of phenolic acids than cellulolytic bacteria, presumably because these phenolics are linked to the hemicellulose fraction of the cell walls and hemicellulolytic bacteria are adapted to these phenolics.

Experiments with some forages have shown a negative correlation between the phenolic constituents of cell walls and apparent digestibility; removal of phenolic acids from cell walls increases digestibility. Ford and Elliott, however, found no relationship between the concentration of any cell-wall constituent and degradability. These latter authors suggested that variability in biodegradability of cell walls is more probably a result of structural features such as cross linking between polymers than concentration of any particular cell-wall constituent.

#### *Effect of Polyphenolics on Animal Performance*

Barry *et al* examined the effects of tannins on nutrient utilisation in sheep fed high-condensed-tannin Lotus. They observed increased levels of growth hormone and their results suggested an increase in the ratio of lipolysis to lipogenesis. Similar trends were reported by Purchas and Keogh, who found lower carcass fat levels in lambs grazing Lotus than in those grazing white clover. This could be due to dilution of fat content by increased N retention, while increased lipolysis may have been mediated by increased secretion of growth hormone.

## Problems of Crop Residues Feeding

### *Post-harvest and Pre-feeding*

Decisions on whether or not to conserve crop residues for feed have to be taken soon after harvesting and often long before feeding them. Lack of convincing economic evidence in favour of their greater use as feed is undoubtedly a restraining factor. Animal scientists are partly to blame in that they have generally emphasised biological rather than economic responses to upgrading and supplementing crop residues. A problem which has a bearing on this is the difficulty of accurately predicting the nutritive (and therefore economic) value of crop residues from simple laboratory techniques, as evidenced by the recent EEC workshop.

The problem is likely to be even greater for tropical crop residues containing anti-nutritional phenolic compounds, especially if feeding systems allow the animals to exercise selective feeding. Cereal straws/ stovers generally either are left in the field or accumulate where the crop is threshed. This is often far from where animals are kept and either the animals must be brought to the field for stubble grazing, or crop residues have to be transported to the animals. The bulk of straws and stovers and lack of transport discourage greater use of straws and stovers as feed. Transporting crop residues, even over short distances, may be uneconomic for small farmers.

The handling and storing of crop residues have been discussed by Hilmersen *et al.* More research and development is required to alleviate problems associated with storage of crop residues. These include risk of loss due to fire, and reduction in nutritive value due to moulding (especially in humid conditions) and damage by vermin and insects. Straws and stovers comprise stem and leaf plus leaf sheath (approximately 1:1 in barley straw), and harvesting, handling and storing systems should minimise the loss of the more nutritious leaf and leaf sheath. In this regard delayed harvesting, or relay harvesting in an intercropped field, would be expected to cause greater loss of leaf and leaf sheath, with a consequent reduction in nutritive value.

Crop residues are characteristically low in metabolisable energy and nitrogen content and thus intake is low. Methods of upgrading straws as feed are well documented and guidelines on researching the subject are available. There has been more emphasis on upgrading straws for cattle than for sheep but goats have received little attention. Upgrading straws for feed is regarded as inappropriate for developing countries, especially for small-scale farmers, because it is expensive and needs technical expertise.

Greenhalgh concluded that in many situations chemical upgrading will be superceded by breeding more nutritious straw, improved harvesting methods and judicious supplementation. The Reading experiments reported here suggest another approach, namely increasing intake of digestible nutrients and therefore animal productivity from crop residues by allowing selective feeding by goats and sheep. The literature on feeding straws to sheep and goats involves experiments where intake and digestibility have been measured under ad libitum feeding. 'Ad libitum' is defined as offering sufficient feed (usually in chopped form) to ensure that 15 to 20% is left (refused) by the end of the feeding period. This approach is standard and has the advantage of minimising selective feeding.

# 4

# Upgradation of Crop Residues

As parts of plants, crop residues contain five different tissue types:

(a) vascular bundles containing phloem and xylem cells;

(b) parenchyma bundle sheaths surrounding the vascular tissue;

(c) sclerenchyma patches connecting the vascular bundles to the epidermis;

(d) mesophyll cells between the vascular bundles and epidermal layers; and

(e) a single layer of epidermal cells covered by a protective cuticle on the outside.

These tissues are digested to different degrees in the rumen. In general, the extent of tissue digestion by ruminal bacteria is as follows: mesophyll and phloem > epidermis and parenchyma sheath > sclerenchyma > lignified vascular tissue. These differences in tissue digestion explain the wide range in nutritive value of crop residues compared to conventional feeds. The cells have two major components: contents and walls. The cell content fraction contains most of the organic acids, soluble carbohydrates, CP, fats and soluble ash. The cell wall fraction includes hemicellulose, cellulose, lignin, cutin and silica. In most crop residues, the cell wall fraction accounts for 60-80 percent of dry matter (DM).

## Cell Walls of Crop Residues

Cell walls of crop residues consist mainly of polysaccharides, protein and lignin. These substances, with small amounts of other components, like acetyl groups and phenols, are organised in a complex three-dimensional structure. Other wall components include suberin, cutin, tannins, waxes and minerals.

### *Polysaccharides*

Major polysaccharides in primary cell walls of most higher plants include cellulose, xyloglucan and pectic polysaccharides, while secondary cell walls contain mainly cellulose and xylans. Cellulose is a highly ordered linear homopolymer of glucose linked by b-1,4-bonds. In all higher plants, cellulose in primary and secondary walls exists in the form of microfibrils. The crystallinity of cellulose microfibrils is highly variable depending on the source and age of the tissue.

In primary walls, the crystallinity has been estimated to be 20-30 percent, while in secondary walls it is 40-70 percent. The crystallinity of the cellulose in wheat internode walls is 30-50 percent and does not change markedly with maturity. Cellulose in mesophyll cell walls from ryegrass leaves, although fibrillar in appearance, had a low crystallinity index ([a]1) compared with >12 for cotton mesophyll cells and 4-5 for non-mesophyll cells. Cellulose molecules in primary walls are heterogeneous in their degree of polymerisation (DP), between 2-6 000, but in secondary walls they are longer and more homogeneous (DP = 14 000).

Cellulose from maize stalk and wheat straw has DP values of 6-7 000. Hemicelluloses are a wide group of polysaccharides that basically share only the property of being soluble in dilute alkali and being able to bind to cellulose by multiple hydrogen bonds and to bind to lignin by covalent bonds. In grasses, the main fraction of hemicellulose is xylans, with a backbone of 4-linked xylose residues and short side chains of arabinose, glucuronic acid and 4-O-methyl-glucuronic acid residues. Most of xylose residues in higher plants are acetylated, mainly on the C-2 hydroxyl groups, but also on C-3.

Hemicellulose polysaccharide concentrations in grasses can range anywhere from 150 to 400 g/kg DM, whereas in legumes, the concentration is much lower, generally between 80-150 g/kg DM. For both grasses and legumes, xylose usually comprises half or more of total sugars of hemicellulosic fraction. Furthermore, rhamnose only exists in the hemicellulosic polysaccharides of legumes.

Pectic polysaccharides are present in the primary cell walls of all seed bearing plants and are located particularly in the middle lamella. They are the major components of the primary cell walls of dicotyledons but account for relatively less of the primary walls in monocotyledons. Three pectic polysaccharides have been structurally characterised from the primary walls of both monocotyledons and dicotyledons: rhamnogalacturonan I, rhamnogalacturonan II, and homogalacturonan. Pectic polysaccharide concentration is quite low in grasses (monocotyledons), generally <10 to 40 g/kg DM, while fairly high in legumes (dicotyledons) ranging from 50 to 100 g/kg. The distribution of the different pectic polysaccharides within the cell walls is largely unknown.

### Proteins

Proteins make up 2 to 10 percent of the primary cell wall of many dicotyledons and some monocotyledons, and may become cross-linked by the formation of isodityrosine or dityrosine. Cell wall proteins may also be involved in covalent bonding with polysaccharides. Glycoproteins seem to be invariably found in primary cell walls. Apparent covalent protein-lignin linkages have also been observed in wheat internodes.

Of the several types of structural proteins known, the best-characterised are the family of hydroxyproline-rich proteins, or extensin. These glycoproteins with rod-like conformations are components of the wall matrix in dicotyledons and in grass walls. Other wall proteins, e.g. glycine-rich proteins, have been found in walls of herbaceous dicotyledons.

### Lignin

Lignin represents between 5-20 percent of crop residue DM. Lignin is described as three-dimensional networks of phenylpropane units. It is generally recognised that the precursors of these building stones are coniferyl, sinapyl, and *p*-coumaryl alcohols, which are transformed into lignin by a complex dehydrogenative polymerisation process. These three aromatic monomers in lignin are referred to as *p*-hydroxyphenyl, guaiacyl and syringyl residues, respectively. Depending upon the number and type of functional groups on the aromatic rings and propane side chains, lignin has variable solubilities.

Wheat straw lignin has higher alkali solubility than wood lignin. When wheat straw lignin is methylated with diazomethane, the number of free phenolic hydroxyl groups in the guaiacyl monomeric units resembled that in pine lignin. Grass lignin is esterified by cinnamic acids, chiefly *p*-coumaric acid through hydroxyls on its monomers. In addition, ether-linked ferulic acids have been observed in lignin from maize stalks, wheat straw, rice straw and bagasse.

Lignin in plant cell walls is physically and chemically associated with wall polysaccharides and proteins. The association between lignin and polysaccharides includes glycosidic linkages, ether cross-linkages, ester cross-linkages and cinnamic acid bridges. On the association between lignin and proteins in straws, limited information is available. Only a covalent protein-lignin linkage was reported in wheat internodes. The strong linkage between lignin and polysaccharides or proteins would definitely prevent cell wall components from enzymatic hydrolysis by ruminal micro-organisms, and thus limit the digestion of cell walls.

### *Other Components*

Other components—including cutin, suberin, tannins, waxes and minerals—are also found in the cell walls. Cutin and waxes are attached to the epidermal walls on plants surface. Cutin is a three-dimensional polyester composed of w-hydroxy and mid-chain hydroxy fatty acids. It is often esterified with phenolic acids, and maintains a close association with pectin in the epidermal cell walls. Cutin appears to be embedded in wax and pectin; these components serve as diffusional barriers that impede ruminal digestion of the intact tissue.

Suberin is a functional component of cell walls. The polyesters that appear in suberised tissue can be esterified with phenolic monomers, oligomers or lignin. Silicon is an important inorganic element in plant cell walls and mainly present in the form of silica in the walls of epidermal cells and leaf hairs. The presence of silica in the cell wall of rice straw can limit the rumen digestibility of polysaccharides. Tannins are phenolic compounds synthesised by some plants as a defence. They may inhibit the activity of specific enzymes, such as cellulases. Since tannins are often insoluble, they can contaminate the crude lignin, resulting in higher analytical value. As a result of complexes with protein, tannins would depress its utilisation, but may not affect cell wall carbohydrates.

### Nutritive Value of Crop Residues

Nutritive value is generally determined by feed composition, intake and utilisation efficiency of digested matter. Thus, the value of a feed depends on chemical composition, digestibility, intake and efficiency. Table 1 contains the nutrient content of some cereal crop residues in China. All of these residues, with the exception of peanut hay, have insufficient CP for efficient rumen fermentation (<9 percent of DM). Apart from sweet potato vines, all have a high crude fibre (CF) or cell wall content (in terms of neutral detergent fibre (NDF)) and low available energy (e.g. NEL). Such high fibre contents are believed to be negatively correlated with voluntary intake, rate of organic matter fermentation, microbial cell yield per unit organic matter fermented, and propionate: acetate ratio in fermentation end products.

Crop residues also have a low mineral content, especially P, and are deficient in vitamins. Therefore, supplementation of crop residues before feeding is necessary, in addition to various treatments. In rice straw, another unwanted constituent is oxalic acid, which can cause rumen disorders. However, oxalic acid can be eliminated by alkali treatment, washing or ensiling. Various crop residues have their own nutritional values and are used for different animal species. Sweet potato vines and peanut hay are relatively rich in protein, available energy and vitamins, and are mainly fed to pigs in most rural areas.

*Table 1. Nutrient content of some crop residues*

| Crop residue | | Analysis on DM basis | | | | | | | |
|---|---|---|---|---|---|---|---|---|---|
| | | DM (%) | $NE_L$ (MJ/kg) | CP (%) | EE (%) | CF (%) | CW (%) | Ca (%) | P (%) |
| Wheat straw | (Ningxia) | 91.6 | 3.27 | 3.1 | 1.3 | 44.7 | 73.0 | 0.28 | 0.03 |
| Maize stovers | (Jiangsu) | 91.8 | 5.23 | 6.5 | 2.7 | 26.2 | 70.4 | 0.43 | 0.25 |
| Rice straw | (Fujian) | 83.3 | 4.11 | 3.7 | 1.6 | 31.0 | 64.4 | 0.11 | 0.05 |
| Sorghum stovers | (Liaoning) | 95.2 | 4.69 | 3.9 | 1.3 | 35.6 | 74.8 | 0.35 | 0.21 |
| Barley straw | (Xinjiang) | 88.4 | 3.69 | 5.5 | 3.2 | 38.2 | 80.1 | 0.06 | 0.07 |
| Soybean straw | (Jilin) | 89.7 | 3.85 | 3.6 | 0.5 | 52.1 | 74.0 | 0.68 | 0.03 |
| Oat straw | (Hebei) | 93.0 | 4.52 | 7.0 | 2.4 | 28.4 | 72.3 | 0.18 | 0.01 |
| Millet straw | (Heilongjiang) | 90.7 | 4.61 | 5.0 | 1.3 | 35.9 | 74.8 | 0.37 | 0.03 |
| Peanut hay | (Shandong) | 90.0 | 5.70 | 12.0 | 2.7 | 24.6 | 88.8 | 0.13 | 0.01 |
| Sweet potato vine | (Yunan) | 91.7 | 5.53 | 8.4 | 2.6 | 19.8 | 36.6 | 1.47 | 0.48 |

According to a survey conducted by Zhou Meiqing, partially feeding pigs with fibrous feeds, including peanut hay and/or sweet potato vines, fresh, dried or ensiled, is a popular practice in Sichuan province, where close to 100 million swine are marketed each year. In this feeding system, fibrous feeds can meet a large percentage of the nutrient requirements: 40 percent of digestible energy, 70 percent of proteins and 80 percent of minerals and vitamins. Wheat straw and rice straw have high contents of cell walls, and are basically used for feeding ruminants. Millet straw and soybean straw, in contrast, are fairly palatable for herbivores, and are mostly used as feed sources for horses, donkeys, mules and rabbits.

Considerable information on DM digestibility of crop residues was obtained from universities and research institutes in China. From the data presented in Table 2, it is clear that *in situ* DM and NDF digestibilities of wheat straw and rice straw are lower than in other residues. The relatively high lignin content in those residues is probably responsible, at least to some extent, for the limited cell wall digestibility.

*Table 2. In situ DM and NDF digestibility (%; 48 hour) of some crop residues*

| *Crop residue* | *ISDMD* | *ISNDFD* |
|---|---|---|
| Wheat straw | 47.6 | 42.2 |
| Rice straw | 42.9 | 40.9 |
| Maize stover | 50.6 | 46.9 |
| Peanut straw | 77.2 | 54.6 |
| Barley straw | 44.8 | - |

Since the 1950s, many *in vivo* digestibility experiments and feeding trials have been conducted in China to determine nutritive values of crop residues. Some results have been summarised elsewhere. It has been concluded that without treatment or nutrient supplementation, feeding most crop residues can just, or barely, meet maintenance energy requirements.

### Effects of Factors on Nutritive Value

A variety of factors have been identified that may influence nutritive value of crop residues. Factors can be divided into three categories: plant, animal and environmental.

#### Plant Factors

The lignin fraction and associated phenolic compounds

are factors most consistently associated with the rigid structure of plants and limited accessibility. The association of lignin with cell wall polysaccharides is also believed to be responsible for resistance of plant cell walls to microbial digestion in the rumen. Table 3 shows the main chemical composition and *in vitro* DM digestibility of three major crop residues widely used in China.

*Table 3. Composition and in vitro DM digestibility of three major crop residues*

| Residue | DM (%) | Chemical composition (% of DM) | | | | | | | IVDMD (%) |
|---|---|---|---|---|---|---|---|---|---|
| | | CP | NDF | NDS | ADF | CEL | HC | ADL | |
| Rice straw | 90.6 | 4.7 | 67.2 | 32.8 | 46.3 | 33.8 | 20.9 | 5.2 | 42.2 |
| Wheat straw | 90.3 | 4.4 | 79.1 | 20.9 | 54.9 | 43.2 | 24.2 | 7.9 | 43.0 |
| Maize stover | 96.1 | 9.3 | 71.2 | 28.8 | 38.2 | 32.9 | 32.5 | 4.6 | 49.1 |

Wheat straw has higher lignin, and therefore lower DM digestibility, compared with maize stover. Although rice straw has a medium lignin content, its DM digestibility is rather low, which may be caused by its relatively high silica concentration. Other plant factors include species, stage of maturity at harvest, cultivar, and proportions of leaf, sheath and stem. All these are believed to influence the nutritive value of crop residues. As plants mature, nutrient digestibility generally declines, linked to a decrease in the digestibility of cell wall components. Xing Tingxian reported that, at an early growth stage, *in vitro* DM digestibility (IVDMD) of wheat straw is pretty high.

As wheat matures, however, the IVDMD of straw progressively decreases. When the grain is completely mature at harvest, the straw has its lowest IVDMD value, resulting from decreased nitrogen content and increased lignification. Xing Tingxian has also reported a variation in chemical composition and digestibility of crop residues among cultivars. Table 4 clearly shows differences in chemical composition, especially in CP, among different rice cultivars, that result in large variations in IVDMD values as crop residues.

*Table 4. Composition and in vitro DM digestibility of straw of various rice cultivars*

| *Residue* | *DM (%)* | *Chemical composition (% of DM)* | | | | | | | *IVDMD (%)* |
|---|---|---|---|---|---|---|---|---|---|
| | | CP | NDF | NDS | ADF | CEL | HC | ADL | |
| Z802 | 89.1 | 3.8 | 61.7 | 38.3 | 42.0 | 32.0 | 19.7 | 3.9 | 48.5 |
| XZ4 | 88.6 | 5.0 | 61.9 | 38.1 | 41.0 | 32.1 | 20.9 | 4.6 | 55.4 |
| V49 | 86.3 | 5.5 | 60.3 | 37.2 | 38.8 | 30.7 | 21.6 | 3.8 | 47.1 |
| V35 | 89.5 | 3.7 | 64.8 | 35.2 | 46.3 | 34.8 | 18.5 | 3.7 | 39.3 |
| Mean | 88.4 | 4.5 | 62.2 | 37.2 | 42.0 | 32.4 | 20.2 | 4.0 | 47.6 |

*Table 5. Composition and in vitro DM digestibility of wheat and rice straw fractions*

| *Fraction* | *n* | *Chemical composition (% of DM)* | | | | | *IVOMD (%)* |
|---|---|---|---|---|---|---|---|
| | | NDF | ADF | CEL | HC | ADL | |
| | | | | *Wheat straw* | | | |
| Whole plant | 16 | 79.1 | 54.9 | 43.2 | 24.2 | 7.9 | 43.0 |
| Stem | 16 | 87.1 | 55.1 | 46.4 | 32.1 | 8.6 | 24.8 |
| Leaf sheath | 16 | 82.4 | 56.0 | 49.1 | 26.4 | 7.0 | 44.5 |
| Leaf blade | 16 | 71.6 | 52.8 | 46.9 | 18.8 | 5.9 | 61.5 |
| | | | | *Rice straw* | | | |
| Whole plant | 16 | 67.2 | 46.3 | 33.8 | 20.9 | 5.2 | 35.7 |
| Stem | 16 | 61.1 | 43.6 | 35.4 | 21.3 | 4.5 | 51.8 |
| Leaf sheath | 16 | 71.9 | 48.4 | 34.6 | 23.5 | 5.5 | 25.4 |
| Leaf blade | 16 | 61.1 | 39.8 | 25.5 | 21.2 | 5.5 | 33.4 |

Another large variation in chemical composition and digestibility is found among straw fractions. Research data from Xing Tingxian showed that stems from wheat straw have a much lower organic matter (OM) digestibility, compared with leaf blades and sheathes (Table 5). This low digestibility can be attributed to their high content of NDF and lignin. In rice straw, the OM digestibility of stems is much higher than that of leaf sheathes and blades (Table 5). Similar results were also reported by Aman and Nordvist.

#### Animal Factors

Little information is available about animal factors that influence the nutritive value of crop residues. Farmers in China have long known that different breeds and types of animals use crop residues with various efficiencies. Cattle, which retain fibrous matter in the rumen slightly longer than sheep or goats, presumably have an advantage with lower quality crop residues. Cross-bred Brahman (*Bos indicus*) steers, when fed hay with 730 g/kg NDF, digested more NDF in the rumen and had longer ruminal retention time for lignin than did Hereford (*B. taurus*) steers.

With increasing popularisation of cross-breeding techniques in China, farmers noted that, on a high concentrate diet basis, hybrid beef steers have much better growth performance than do native breeds; in contrast, on a low-quality fibrous feed, the contrary was observed. Although the exact mechanism of this difference in animal performance between hybrid and native breeds is unclear, an inherent difference in food intake and digestion capacity may be responsible.

#### Environmental Factors

Some environmental factors, including location, climate, soil fertility and soil type, seem to influence the nutritive value of crop residues. Recent studies have indicated that there can be significant differences in chemical composition and digestibility of crop residues grown on different soil types (Table 6). Irrespective of crop cultivar, straw from wheat grown in the so-called tide soils had considerably higher CP content and lower fibre (NDF, ADF and ADL) content than straw from drab soils. These are probably the cause of digestibility differences.

*Table 6. Composition and in vitro DM digestibility of wheat straw by soil type*

| Parameter | Soil type | Wheat cultivar | | |
|---|---|---|---|---|
| | | No. 3039 | Bao feng | ZY No.1 |
| CP | Tide soil | 6.4 | 4.8 | 4.6 |
| | Drab soil | 3.7 | 3.3 | 3.3 |
| NDF | Tide soil | 64.4 | 71.2 | 58.5 |
| | Drab soil | 69.7 | 74.8 | 75.9 |
| ADF | Tide soil | 48.6 | 57.3 | 45.7 |
| | Drab soil | 58.4 | 55.1 | 64.6 |
| ADL | Tide soil | 7.4 | 8.3 | 6.2 |
| | Drab soil | 7.4 | 8.5 | 9.6 |
| IVDMD | Tide soil | 48.6 | 49.4 | 54.9 |
| | Drab soil | 53.6 | 45.3 | 33.8 |

## Methods of Processing and Feeding

Ruminants despite their unique and highly efficient digestive system, are not able to extract sufficient energy to grow and produce milk from low quality or highly lignified residues. These must be properly processed or treated in some way to make them useful for production.

### Traditional Processing

Historically, many fibrous crop by-products have been used as energy sources for ruminants in China. More than 1 000 years ago, during the Song Dynasty, people used the following method of processing and feeding crop residues: Mix finely chopped straw with wheat or millet bran and beans, slightly soak them with water, and then feed animals *ad libitum*.

These ancient processing and feeding methods include particle reduction and reconstitution of roughage, and are still included in university textbooks and scientific publications. Chopping and water soaking are popular practices for crop residue feeding throughout the country. Although they do not always result in consistent improvements in animal performance, they definitely result in reduced diet wastage and diet selection. A method called *Jiegan Nian Qing* has been widely used in the southern part of Shanxi Province.

A thick layer of wheat or rice straw is spread on a flat threshing ground. A layer of freshly cut alfalfa is put on top, followed by another layer of straw. A heavy stone roller is passed over the layers, squeezing out the alfalfa juice, which is absorbed by the straw. The alfalfa treated in this way can be dried much faster, with minimum leaf loss, and at the same time, the alfalfa juice absorbed by the straw enhances the straw feed value. This method is very useful in those areas where alfalfa drying is difficult.

### Physical Treatment

Numerous physical processing techniques to enhance the utilisation of crop residues by ruminants have been used, with varying degrees of success. Common methods are— including grinding and pelleting, irradiation and steam treatment -will be briefly reviewed as they relate to crop residue utilisation in China.

#### Grinding and pelleting

The most studied physical treatments for enhancing crop

residue use by ruminants are grinding and pelleting. Grinding, or fine chopping, decreases particle size, increases surface area and bulk density of both leaf and stem fractions, and hence raises rumen microbial accessibility or feed intake. The increase in intake due to grinding is generally higher with low quality than with high quality residues, and with small and young animals rather than with older and larger animals. The critical feed particle size to exit the rumen is smaller in sheep than in adult cattle and therefore a greater degree of grinding is necessary before they leave the rumen.

Screen sizes for hammer mill grinding in China range from 2.5 mm to 25 mm. Considering differences in intake between animal species and the energy expenditure for grinding, Xiong recommended 6 mm for sheep and 12 mm for cattle as the appropriate screen sizes for hammer mills. Ground crop residues are often pelleted or cubed before feeding. Benefits derived from pelleting include a further increase in density, decreased dustiness and easier handling. However, DM digestibility of pelleted straws is depressed relative to the long or chopped forms, primarily due to faster passage rate.

Pelleting usually augments straw intake due to quicker passage, which can offset the negative effect from decreased digestibility. Therefore, the net benefit of feeding pelleted crop residues in practice is increased energy intake and animal performance. In dairy cows, fine grinding and pelleting of forages can dramatically reduce rumination and rumen digestion times. Consequently, saliva production is reduced and the rumen fermentation pattern is altered, together with reduction in acetate/propionate ratio. This is believed to be the reason for the milk fat reduction with ground and pelleted forages.

### *Kneading*

As milk fat can be lowered with finely ground or pelleted straws, the development of another physical processing method was necessary. Recently, a novel method for processing crop residue using a kneading machine has been reported. When fibrous crop residues enter the machine, they are kneaded into threadlike fibres or hairs with no apparent stem internode structure. Kneading extensively destroys the rigid structure and thus significantly increases voluntary intake. Unlike other physical processing, such as grinding or pelleting, rubbing of crop residues produces long threadlike fibres, and therefore should not affect milk fat content.

Compared to chopping, kneading requires higher energy expenditure. Several studies have been conducted to compare kneading with traditional chopping. Sun Zhongyin *et al.* reported that dairy cows fed with scrubbed soybean residue had higher dietary DM intake and milk production than with untreated residue. A similar result with kneaded maize stover fed to dairy cows was reported by Zhao Xiyou and Sun Qinglin. Kneading treatment is becoming popular in China.

### *Irradiation*

Irradiation treatment of lignocellulosic materials to improve the utilisation of cell wall polysaccharides dates back to the work of Lawton *et al.* They found that when basswood was irradiated with high velocity electrons, rumen bacteria fermentation was increased. Electron irradiation of straw can also increase polysaccharide digestibility by ruminal micro-organisms. Based on volatile fatty acid (VFA) production during fermentation, the optimum dose appears to be at 2.5 ¥ 108 rad. *In vitro* DM disappearance increased with irradiation dosage up to 108 rad. Several studies on irradiation of crop residues for increasing their nutritive value have been conducted in China.

Meng Qingxiang and Xiong Yiqiang treated wheat straw with a combination of g-rays from a cobalt-60 source and $NH_3$ or $NH_3$ plus CaOH at different moisture levels. They found that irradiation doses had a significant interaction with the moisture level (20, 40 or 60 percent for ammoniation, and 40, 50 or 60 percent for $NH_3$ + CaOH). On either chemical treatment, as irradiation and moisture level increased, *in situ* DM disappearance (ISDMD) increased and NDF content diminished dramatically. These results suggest that responses to irradiation at a lower dosage can be compensated by higher moisture levels.

In another study, Gu Chuipeng *et al.* found lower contents of fibrous fractions and elevated ISDMD with irradiation of rice straw. When straw was irradiated at dosage of 0, 5 ¥ 106, 107, 5 ¥ 107 and 108 rad, the ISDMD were 54.0, 54.7, 57.5, 71.0 and 75.5 percent, respectively. Treatment of rice straw with a combination of electron irradiation and NaOH also resulted in a considerable higher glucose release. Although irradiation is very effective in improving the nutritional value of fibrous crop residues, it remains unfeasible at the farm level.

### *High pressure steaming*

High pressure steaming to improve the nutritive value of low quality feeds was closely studied at the Inner Mongolian Academy of Animal Sciences. Ground wheat straw or other crop residues are placed in a closed steel tank and saturated with high pressure steam. When the expected temperature and time are reached, a tank valve is suddenly opened allowing materials to enter a pressure-release tank through a specially designed tube. This high pressure steaming and explosion result in a brown straw with looser structure. High pressure steaming markedly decreases straw CF (Table 7) and therefore increases the *in vitro* DM digestibility (Table 8).

*Table 7. Composition (% of DM) of wheat straw after high-pressure steaming*

| | DM | CP | CF | Ash |
|---|---|---|---|---|
| Before steaming | 91.7 | 2.57 | 43.5 | 4.4 |
| After steaming | 94.8 | 2.84 | 37.8 | 9.1 |

*Table 8. In vitro digestibility (%) of fibrous feeds after high-pressure steaming*

| | WS | CS | RS | RS | SD |
|---|---|---|---|---|---|
| Before steaming | 38.7 | 52.1 | 40.1 | 40.3 | 24.9 |
| After steaming | 55.5 | 75.5 | 59.6 | 52.7 | 43.3 |

Results from an *in situ* study showed that NDF digestibility of the treated wheat straw was increased by 68 percent in rumen-fistulated sheep and by 233 percent with caecum-fistulated pigs. Rumen VFA concentration was also increased 9.9 percent in sheep fed diets based on the high pressure steamed wheat straw, compared to untreated straw. In lamb feeding trials, animals were fed equal amounts of mixed concentrate and wheat straw per day per animal. Lambs fed high pressure steamed straw ate more of it and gained faster than lambs with untreated straw.

Compared with chemical treatment, high-pressure steam does not require reagents and thus minimises potential environmental pollution. In relation to other physical treatments, high-pressure steam is more effective in improving crop residue nutritive value. However, it implies high investment for equipment and a steam generator, and it has not been developed for wider utilisation throughout the country.

### Chemical Treatment

Since the beginning of the 19th century, attempts have been made to improve the digestibility and nutritive value of crop residues. A major breakthrough was chemical treatment to remove encrusting substances. Many chemicals have been screened in laboratory experiments for their potential to enhance digestibility. However, only three are being routinely used in animal research: sodium hydroxide (NaOH), ammonia ($NH_3$), and calcium hydroxide (CaOH). The modes of action of chemical treatment on crop residues, especially with alkalis, have been described by Klopfenstein:

(1) hemicellulose solubilisation,

(2) increases in cellulose and hemicellulose digestion, and

(3) increases in digestion rate for cellulose and hemicellulose.

The data on ammoniation of maize stover and rice straw from many studies support these modes of action for the chemical treatment.

#### *Sodium hydroxide treatment*

Sodium hydroxide treatment of crop residues has been investigated and used in some areas of the country since the late 1970s. The procedure basically followed the "dry method," where NaOH is applied at 3-5 percent and the moisture content is 20-30 percent of DM. Alkali treatment may saponify the ester bonds between lignin and carbohydrates or the phenolic acid-carbohydrate complexes in plant cell wall. Through these effects, structural carbohydrates in both lignified and unlignified plant tissues become more digestible, with consequent increases in rate and digestibility.

The treatment with NaOH results in increases in crop residue palatability and digestibility, and in animal performance. Steers fed rations based on NaOH-treated wheat straw gained 20 percent faster than did the control group when concentrate was half of total ration. Ye Risong *et al.* reported that dairy cows fed NaOH-treated rice straw diets ingested 1.9 kg more straw and produced 1.4 kg more milk per day than those on untreated-straw diets. Although NaOH treatment works effectively in improving the nutritive value of crop residues, NaOH is expensive, corrosive and its use may result in significant excretion of sodium ions in animal excreta. Long-term accumulation of sodium may lead to soil fertility problems and environmental pollution. Thus, application of NaOH treatment of crop residues is not popular with the farmers at present.

#### *Ammoniation*

Since the middle of 1980s, ammoniation of crop residues has drawn a great deal of attention in China due to several advantages: effectiveness in improving digestibility, addition of non-protein nitrogen to treated residues, and absence of sodium accumulation in soils. Most data have shown a decreased NDF content, but little change in ADF and ADL contents of crop residues due to ammoniation. The results suggest that ammoniation can break the linkage between hemicellulose and lignin and make the hemicellulose fraction partially soluble to NDF solution.

The soluble hemicellulose would be highly digestible by ruminal micro-organisms. After being ammoniated, treated crop residues have an increased N content relative to untreated residues (Table 9). Many *in vitro* and *in vivo* digestion trials have been conducted to evaluate the effect of ammoniation on digestibility of different crop residues. Ten studies with ammoniated crop residues indicated an average increase in digestibility of 24.3 percent (12.4-44.6) or 11.2 (6.4-17.8) percentage units. Animals ingested ammoniated residues faster than the untreated (Table 10). Improvements in the feeding value of crop residues due to ammoniation have been observed in many feeding trials. Some of the results are summarised in Table 11.

*Table 9. Effect of ammoniation on composition (% of DM) of crop residues*

| CP | | NDF | | ADF | | ADL | |
|---|---|---|---|---|---|---|---|
| UNT | AMM | UNT | AMM | UNT | AMM | UNT | AMM |
| | | | *Wheat straw* | | | | |
| 3.5 | 9.1 | 89.1 | 78.9 | 53.7 | 54.2 | 14.3 | 14.1 |
| 2.8 | 6.6 | 78.1 | 69.2 | 51.3 | 51.3 | 19.7 | 14.5 |
| - | - | 84.2 | 76.4 | 50.3 | 49.9 | 15.1 | 14.6 |
| 3.3 | 9.6 | - | - | - | - | 18.7 | 17.1 |
| 4.2 | 4.5 | 77.7 | 75.9 | 51.3 | 49.4 | 10.1 | 9.9 |
| | | | *Rice straw* | | | | |
| - | - | 69.7 | 65.9 | 51.2 | 47.2 | 12.0 | 10.4 |
| 6.1 | 13.5 | 75.0 | 71.4 | 49.9 | 48.7 | 8.7 | 8.3 |
| 5.6 | 10.1 | - | - | - | - | 11.3 | 9.8 |
| 3.7 | 7.3 | 67.2 | 65.0 | 48.6 | 45.1 | 5.1 | 5.2 |
| | | | *Maize stover* | | | | |
| 6.9 | 11.0 | - | - | - | - | - | - |
| 10.7 | 27.2 | 69.4 | 63.9 | 38.7 | 36.2 | 4.7 | 4.6 |

Computer simulation results on beef production under different practical conditions were always in favour of ammoniated wheat straw over untreated straw when comparison was made on the basis of maximum benefit per unit of body weight gain.

*Table 10. Effect of ammoniation on ingestion rate of crop residues*

| *Residue* | *Treatment* | *Animal* | *Intake rate* |
|---|---|---|---|
| | Ammoniated | Native steer | 25.7 |
| Wheat straw | Untreated | Native steer | 42.1 |
| | Ammoniated | Crossbred cattle | 59.0 |
| | Untreated | Crossbred cattle | 94.0 |
| Rice straw | Ammoniated | Dairy heifer | 25.7 |
| | Untreated | Dairy heifer | 44.0 |
| | Ammoniated | Native steer | 20.2 |
| Maize stover | Untreated | Native steer | 23.5 |
| | Ammoniated | Crossbred cattle | 49.4 |
| | Untreated | Crossbred cattle | 68.4 |
| Soy straw | Ammoniated | Goat | 42.0 |
| | Untreated | Goat | 45.0 |

The maximum benefit is expressed as minimum concentrate consumption per unit gain, or minimum feed cost per unit gain. Based on the beef market situation at that time, a two-stage feeding optimised system was proposed. During the first period with minimum concentrate consumption per unit gain, cattle should be fed a largish amount of ammoniated crop residues to maintain a relatively low rate of daily gain. During the finishing period with minimum feed cost per unit gain, cattle were to be fed on low crop residue and high concentrate diets to allow faster gain rates.

The computer simulation results from this two-stage feeding system compare well to actual feeding results from Beijing, Shandong, Shanxi and Henan. In another study, Meng and Xiong found that lambs fed ammoniated wheat straw had increased dietary intake, body weight gain and better concentrate conversion efficiency compared with animals fed untreated wheat straw. The magnitude of the improvement gradually declined with increasing proportion of mixed concentrates in the diet.

Regression showed that feeding ammoniated straw diets to lambs could benefit either by increased daily gain at similar concentrate level, or by less concentrate feed consumption at the same rate of gain. Based on the results, it was calculated that each tonne of ammoniated wheat straw replacing untreated straw could produce 105.2 kg (37-159.2 kg) more of liveweight gain at concentrate levels from 22 to 72 percent. When lambs gained at equal rates, each tonne of ammoniated wheat straw could save about 285.4 kg of mixed concentrates or grains at the above range of concentrate levels. This conclusion agrees well with experience from commercial animal production: each tonne of ammoniated crop residues when replacing untreated residues could save 250-300 kg of grain in cattle or sheep.

*Table 11. Effect of ammoniation on lactation performance of dairy cows*

| *Residue* | *Treatment* | *n* | *Milk (kg/d)* | *FCM kg/d* | *Milk fat (%)* |
|---|---|---|---|---|---|
| Wheat straw | Ammoniated | 6 | 21.4 | 21.0 | 3.78 |
| | Untreated | 6 | 18.4 | 18.0 | 3.68 |
| | Ammoniated | 6 | 22.3 | 20.6 | 3.48 |
| | Untreated | 6 | 22.0 | 20.1 | 3.42 |
| Maize stover | Ammoniated | 8 | 20.1 | 18.8 | 3.57 |
| | Untreated | 8 | 17.5 | 16.4 | 3.59 |
| | Ammoniated | 6 | 16.4 | 16.3 | 3.98 |
| | Untreated | 6 | 15.7 | 15.4 | 3.88 |

Many studies have also demonstrated that feeding ammoniated crop residues greatly improved lactating performance of dairy cows. Table 11 summarises the results of 4 trials and shows that ammoniation of crop residues increased actual yield on average by 1.7 kg (20.1 vs 18.4 kg) without changes in milk composition, including fat percentage.

*Other treatments*

Since limestone is available cheaply in China, the use of $Ca(OH)_2$ to treat crop residues attracted a great deal of interest from the 1950s. Calcium hydroxide is generally less effective in treating crop residues than other alkaline sources, such as NaOH or $NH_3$. Combining $Ca(OH)_2$ with urea or other alkalis seems to solve this problem. Combining $Ca(OH)_2$ with urea, Mao Huaming and Feng Yanglian showed that rice and wheat straw treatment increased the CP content by 3.5 times and *in situ* DM digestibility by 69.8 percent.

In a feeding trial, dairy heifers fed such treated rice straw showed significant increases in dietary DM intake, weight gains, feed conversion as compared with those fed the untreated straw. Cao Yufeng *et al.*reported significant improvements in the nutritive value of wheat and rice straws as a result of combination treatment with urea, calcium hydroxide and common salt. Table 12 shows the changes in NDF, ADF, ADL, cellulose and hemicellulose content and *in vitro* DM digestibility before and after treatment.

Combined treatment reduced the content of NDF, ADF and hemicellulose, but did not change the content of cellulose and lignin over the untreated straw. The *in vitro* DM digestibility of treated straws was enhanced relative to untreated straws. Cattle fed the combined diets had somewhat more dietary DM intake, better daily gain, improved feed conversion and considerably reduced feed cost per kg of weight gain than cattle on either untreated or ammoniated straw diet.

*Table 12. Effect of combined treatment on composition (as % of DM) and digestibility (%) of wheat and rice straw*

| | NDF | ADF | CEL | HC | ADL | IVDMD |
|---|---|---|---|---|---|---|
| | | | *Wheat Straw* | | | |
| Untreated | 84.2 | 50.3 | 33.6 | 33.9 | 15.1 | 36.2 |
| Urea + $Ca(OH)_2$ + salt | 74.5 | 47.4 | 32.5 | 27.0 | 14.2 | 43.7 |
| | | | *Rice straw* | | | |
| Untreated | 69.7 | 51.2 | 34.6 | 18.6 | 12.0 | 40.9 |
| Urea + $Ca(OH)_2$ + salt | 61.4 | 45.4 | 30.5 | 16.0 | 9.9 | 51.2 |

Other combination methods for treatment of crop residues with sodium hydroxide and urea, ammoniation and enzyme, and ammoniation and ensilage have also been reported elsewhere. In each case, the nutritive value was improved, but these methods have not so far been taken into practice.

## Biological Approach

### Microbial ensilage

Ensilage of whole fresh maize plants is only practised for large-scale feedlots and dairy farms. For small-scale family farms, ensiling dry crop residues after reconstitution of moisture is usually the best way for preserving feeds, since farmers do not have suitable equipment to quickly harvest their cereal plants. Another reason is that they have to sow promptly the next crop in most regions with a double-cropping system. Ensiling dry crop residues involves actions such as chopping, reconstitution of moisture, pressing and mixing with certain additives, including micro-organisms such as lactic acid producing bacteria, cellulolytic bacteria, for proper fermentation and nutrient preservation.

A large number of dry crop residues have been successfully ensiled with addition of microbial products in China in recent years. This method is commonly called "microbial ensilage," or *Weizhu* in Chinese. Some bacterial products with specialised functions and warranted quality have been developed and approved for practical use by the government. Wu Keqian and Meng Qingxiang *et al.* ensiled wheat straw with addition of a specific microbial product containing bacteria that function as lactic acid and propionic acid producers and cellulose degraders, and fed it to cross-bred steers.

The results showed that microbial ensiling resulted in reduction of NDF, ADF, cellulose and hemicellulose, and an increase in *in situ* DM digestibility (Table 13). In some feeding studies, it was shown that microbial ensilage of crop residues such as wheat straw, rice straw, maize stover or soybean straw caused increased daily gains, feed intake and feed conversion, and decreased feed cost per unit gain in growing ruminants. Several studies also indicated that lactating cows fed diets based on microbial ensiled straw had increased milk and fat-corrected yield, and slightly higher milk fat percentages, compared with diets based on untreated straw.

*Table 13. Composition of wheat straw before and after microbial ensiling*

| *DM (%)* | *NDF* | *Composition ADF* | *CEL* | *(% of DM) HC* | *ADL* | *ISDMD (%)* |
|---|---|---|---|---|---|---|
| Untreated | 87 | 78 | 51 | 32 | 27 | 20 42 |
| Microbial | 33 | 70 | 50 | 33 | 20 | 16 46 |
| Untreated | - | 83 | 60 | 45 | 23 | 15 37 |
| Microbial | - | 79 | 57 | 43 | 21 | 14 41 |

Another significant effect of microbial ensilage of dry crop residues is probably to hydrate and weaken plant structures so that less energy is expended on rumination. Ensiled crop residues usually have good palatability for ruminants, and thus high intake. In comparison with ammoniated straw, microbial ensiled residues give higher intake, faster rate of passage and therefore better performance. Other advantage of microbial ensilage is its low input cost for acquiring microbial products and accessories, e.g. plastic sheets, and therefore microbial ensilage is considered a better method to enhance the feeding value of dry crop residues. However, microbial ensilage generally results in lower digestibility than ammoniation.

Another disadvantage of microbial ensilage includes substantial loss of organic material that would otherwise be rapidly fermented in the rumen. As a result, it is still argued academically whether the anaerobic ensilage of such ready digestible materials within crop residues is economically beneficial to the animal. Further in-depth research is required to select bacteria strains that selectively degrade cell wall fractions, especially lignocelluloses.

#### *Treatment with White Rot fungi*

Because White Rot fungi can effectively attack lignin and cellulose, their use to treat lignocellulosic material to increase digestibility has been studied quite extensively in other countries, but little in China. Xiao Xunjun and Peng Jun at China Agricultural University treated wheat straw with strains of *Cyathus stercoreus, Bjerkandera adusta, Dichomitus squalens, Pleurotus* spp. and *Pleurotus ostreatus* for 30 days, and showed that treatment decreased NDF from 71.4 to 67.4, 59.2, 62.7, 65.0 and 67.9 percent, and

ADF from 53.1 to 50.3, 45.1, 46.0, 50.0 and 51.3 percent, respectively.

After fermentation by the five fungus strains, a considerable loss was found in lignin, from 23 to 44 percent, and in DM, from 11 to 17 percent. There was no apparent loss in cellulose and hemicellulose. When wheat straw was incubated *in vitro* with mixed rumen micro-organisms for 24 hours, DM digestibility was increased 11 and 8 percent for the treatment with *Bjerkandera adusta* and *Phleurtus* spp., respectively, compared with the untreated control. Straw digestibility with the other three strains did not change.

When activity of polysaccharide-degrading enzymes (FPase, avicelase, CMCase, xylanase) and ligninase (Mn-dependent peroxidase) was measured, Peng Jun found that enzyme activity varied considerably with different fungus strains. It was also noted that most White Rot fungi grew slowly on common crop residues and could not effectively compete against other microbes. These observations suggest that effective breakdown of crop residue cell walls by White Rot fungi in practice will require selection or creation of better strains, and also further refinement of the current treatment techniques.

#### *Mushroom-substrate residues*

Crop residues have been used as a substrate to grow mushrooms. This practice is a very profitable business in some areas of the country. The substrate residue after mushroom harvest can be used to feed animals. The most commonly used crop residues are cottonseed hulls, wheat straw, rice straw and maize stover. The residues usually have higher CP and lower CF contents compared with the original substrate. Yang Xunyi *et al.* reported that after the 2nd, 3rd and 4th harvest of mushrooms, the CP content of the residual substrate increased by 32.5, 44.2 and 60.9 percent, while its CF content reduced by 42.4, 48.1 and 50.4 percent, respectively.

When the substrate residue was included in growing pig diets at level of 5 percent, there was no significant difference in average daily gain and feed conversion. However, growth performance of pigs decreased with increased substrate residue inclusion. The only benefit from inclusion of the substrate residues at a low rate in pig diet is the decreased consumption of concentrate or feed cost per unit of body weight gain.

Undoubtedly, the use of crop residues for mushroom production is a very good approach in China's agro-ecosystem. Research data have also indicated that some species or strains of mushrooms have strong enzymatic activities digesting cellulose and lignin. Regarding the feeding value of this residue, however, more work remains to be done before any overall recommendations can be given.

### *Enzymatic treatment*

The use of enzymes to attack the lignocellulose structure of crop residues for enhancing their feeding value has been attractive. Crude enzyme products, with cellulolytic and hemicellulolytic capability, are usually added to fibrous feeds in attempts to improve their digestibility. Wang An observed that treatment of maize stover with an enzyme product, prepared from *Trichoderma viride*, reduced the contents of some cell wall components and enhanced the ruminal digestibility in sheep (Table 14).

Huang Jianhua and Huang Jianhua *et al.* treated maize stover and spent grain from malting (60:40) with an enzyme mixture containing cellulase, proteinase and amylase, and measured the effects on the performance of finishing pigs and laying hens. Inclusion of 10 percent of the treated maize stover and spent grain mixture in the diet did not affect gain rate of pigs or egg production of hens, but reduced by ¥ 0.25 the feed cost per kg of liveweight gain with the pigs and by ¥ 0.49 per kg of egg production with the laying hens.

*Table 14. Effect of cellulase addition to maize stover on fibre fraction content and in situ digestibility in sheep*

| Item | Control | With cellulase |
|---|---|---|
| | *Chemical composition (%)* | |
| NDF | 58.2 | 56.2 |
| ADF | 37.3 | 35.3 |
| ADL | 4.9 | 4.5 |
| CEL | 32.3 | 29.8 |
| HEM | 20.9 | 20.9 |
| | *Digestibility (%)* | |
| DM | 39.8 | 45.8 |
| NDF | 27.4 | 31.3 |
| ADF | 29.4 | 31.7 |
| ADL | 16.0 | 17.8 |
| CEL | 31.5 | 33.7 |
| HEM | 23.6 | 30.5 |

Commercial cellulase products were also added to diets to increase the supply of readily available carbohydrate. When the enzyme products were included at 0.1-0.2 percent of the diet of pigs, cattle and geese, animal performance was considerably improved. Chen Xiafu *et al.* also reported the use of crude enzyme products prepared from *Trichoderma viride* as feed additives for growing rabbits. In eight growth trials, rabbits fed on a diet with addition of the cellulolytic enzymes gained 17.5-39.3 percent faster than the control.

The difference was consistent and highly significant. Although there is a tendency toward increased use of cellulolytic enzymes in animal feeds, at present the cost of suitable enzymes is too high for commercial use. Obviously, advances in biotechnology and increased production of effective enzymes would be expected to lower the cost of enzymatic treatment.

# 5

# Ensiling Crop Residues

Silage must be made from plant material with a suitable moisture content, which ranges from about 55% to 70% depending on the construction of the storage structure and hence the degree of compression and the amount of water that will be lost during storage. For corn, harvest begins when the whole-plant moisture is at a suitable level. For pasture-type crops the grass is mowed and allowed to wilt for a day or so until the moisture drops to an acceptable level. In New Zealand and Northern Europe it is harvested directly from the paddock by a tractor-mounted silage chopper, blown into a trailer-mounted cage and transported to the silo or heap.

In the US, the plant material is collected, chopped into pieces about 1/2" (13 mm) long and packed into the storage. In the early days of mechanised agriculture, cornstalks were cut and collected manually using a knife and horsedrawn wagon, and fed into a stationary machine called a "silo filler" that would chop the stalks and blow them up a narrow tube to the top of a tower silo. Current technology uses mechanical forage harvesters that collect and chop the plant material, and deposit it in trucks or wagons. These forage harvesters can either be units that are pulled by a tractor, or they can be self-propelled machines.

Harvesters blow the silage into the wagon via a chute at the rear or side of the machine. Once a wagon is full, the wagon is taken back to the farm. For tower silos, the silage is emptied into a stationary blower which blows the silage up into the silo. Silage may also be emptied into a bagger, which puts the silage into a large plastic bag that is laid out on the ground. In New Zealand and Northern Europe the silo or 'pit' is often a concrete bunker built on the side of a bank so that chopped grass can be dumped in at the top and fed out from the bottom in winter.

The disadvantage is that it requires considerable effort to compress the stack in the silo to cure properly. So much of the silage made there is stacked in a heap on the ground called a clamp and compressed with a tractor rolling over the top as it is stacked. The result is a fat pancake-like stack. It is much easier to get the cured silage from such stacks in winter when the ground is very wet because the tractor can access all sides of the heap.

### Fermentation of Silage

Silage undergoes anaerobic fermentation, typically beginning about 48 hours after the silo is filled. Traditionally, the fermentation is caused by indigenous microorganisms; today, some silage is inoculated with specific microorganisms to speed the fermentation or to improve the resulting silage. The process converts sugars to acids and exhausts any oxygen present in the crop material. The fermentation is essentially complete in about two weeks.

The fermentation process releases a liquid. The amount of liquid can be excessive if there is too much moisture in the crop when it is ensiled. Silo effluent contains nitric acid ($HNO_3$), making it corrosive. It also can be a contaminant of lakes and streams, since the high nutrient content can lead to eutrophication (growth of algae blooms).

Silage must be protected from oxygen or it will spoil. Silage must be firmly packed to minimise the amount of air present. Silos are hazardous, and people die every year in the process of filling and maintaining them. The machinery used is dangerous, and with tower silos workers can fall from the silo's ladder or work platform. There are also respiratory hazards from the fermentation process itself.

Nitrogen dioxide gas is released in the early stages of fermentation, and can kill. The reduced oxygen environment inside the silo can cause asphyxiation, and molds formed when air is allowed to reach cured silage can cause toxic organic dust syndrome. When filling a silo, the air can also become filled with fine dust particles, which can be ignited by a flame or spark, causing an explosion.

The ensiled product retains a great deal of the nutrients present in the plant, much more so than if the crop were dried and stored as hay or stover. Silage is most often fed to dairy cattle, because they respond well to highly nutritious diets.

## Different Kinds of Silos

### *Tower Silo*

In China, tower silos are constructed from brick, and are several metres in diameter and 10-20 m in height. The advantages of this type of silo include: long life, small space required, low storage losses, and possibility for mechanisation. Both the filling operation and daily extraction can be mechanised. However, tower silos are expensive, and therefore not widely used in China, with the exception of some state-owned farms.

### Cellar Silo

The cellar type is the most common silo on individual farms. Round or square concrete silos are usually built inside houses for protection from the weather. Advantages are lower cost and easy management. Size can be adjusted according to scale of production. Cellar silos are suitable for rural conditions in China. A disadvantage is high effluent loss, especially with clay walls.

### Trench Silo

This type is generally built underground or semi-underground, with two solid walls of 1.5-2 m in height. Advantages are similar to the cellar silo, but the trench silo is more suitable for mechanisation. The tractor can be driven on top from one side to the other for compaction purposes. After compaction, it is covered with a plastic sheet pressed down with soil, sandbags or straw bales to maintain anaerobic conditions. On many dairy farms, trench silos are built on the surface of ground. This type of trench silo resembles a bunker silo, but has vertical walls of 0.4-0.5 m in thickness and 3-4 m in height. This design makes mechanisation more convenient, and may also prevent bottom leakage.

### Stack Silo

This type of silo implies a pile of material on the ground surface. On flat and dry ground, plastic sheet is placed underneath and the material is laid in a stack. The top is covered with plastic and sealed all round with soil. Sandbags or old tyres, or any other suitable objects, are placed on top to prevent the top cover from being blown away by the wind. The advantages of the stack silo are low cost and flexibility of placement.

#### *Plastic Silo*

Animal scientists from Neijiang, Sichuan Province, successfully ensiled sweet potato vines in plastic bags in 1978. The plastic silo is similar to the stack silo but it is covered with plastic sheets of polyvinyl chloride (PVC) or polyethylene. Alternately, the silo can be made in bags with sealed tops. The stack silo is also inexpensive and can be placed anywhere. However, labour requirements are high due to manual filling and handling. A specific machine has recently been developed by the Grassland Institute, Chinese Academy of Agricultural Science, to fill plastic bags with forage.

### Silage Making

#### *Control of Moisture Content*

Ensilage can only be successful, with minimum DM and nutrient losses, when the moisture content of the raw material is kept to a suitable level. Although silage may be made within a large range of moisture contents, DM should be over 20 percent to assure silage quality. There are many disadvantages to ensiling crops with high moisture content.

First, ensiling of wet materials results in the generation of a large volume of effluent, which not only poses disposal problems, but also carries off valuable, highly digestible nutrients in solution. The amount of effluent increases with silo height, due to pressure. Effluent is produced when moisture is above 75 percent.

Secondly, the critical pH value for clostridial growth varies directly with the moisture content of the plant material, and unless soluble carbohydrate levels are exceptionally high, ensiling wet crops will encourage clostridial fermentation, resulting in high losses and reduced nutritive value.

Thirdly, even if the water-soluble carbohydrate (WSC) levels are adequate to ensure lactic fermentation, very wet silages may still be nutritionally undesirable because voluntary DM intake of these is frequently low.

Finally, drier plant materials are preferred because they are easier to handle and a higher quantity of DM can be carried per trailer load.

Moisture content of forage and grasses is above 80 percent when harvested at a suitable stage. Therefore the moisture content should be reduced by field wilting. It takes 4 to 6 hours to wilt in dry regions such as the north western provinces and Inner Mongolia, and 6 to 10 hours in north eastern and northern areas. Longer periods may be needed in southern provinces, depending on climate and weather, but it is not desirable to exceed 24 hours. When weather conditions are unfavourable, field wilting should be avoided to prevent nutrient loss due to rain leaching. In these cases, other methods should be considered.

In contrast, the moisture content of cereal straw is generally too low to allow tight packing, so cereal straw and stover should be finely chopped. Sometimes water should be added to bring moisture content to a suitable level. Sweet potato vines are high in moisture content and wilting is necessary. Chopped vines are usually mixed with finely chopped straw or bran prior to ensiling to increase overall DM content. The moisture content of plant materials may be measured with instruments, but it is usually estimated manually on farm.

Samples of chopped and minced grass or leguminous forage are grasped tightly by hand for one minute or so to estimate moisture content. If juice can be extracted, moisture content is above 75 percent. If the material remains together but without juice, moisture content is between 70-75 percent. If the material has elasticity and spreads out slowly, moisture content is 55-65 percent. If the material spreads out quickly, moisture content is about 55 percent. If the material breaks, moisture content may be below 55 percent.

#### *Chopping, Compaction and Sealing*

Prior to ensiling, plant materials should be chopped. The fineness of chopping varies with moisture content and nature of the material. The following guidelines can be used, but, in principle, rough and hard materials should be finely chopped, while delicate and soft materials can be roughly chopped.

| | |
|---|---|
| High moisture forage (Moisture >75%) | chop to 6.5-25 mm |
| Wilted materials (Moisture 60-70%) | chop to 6.5 mm |
| Whole maize plant | chop to 6.5-13 mm |

Some advantages of chopping are:

- Chopping facilitates compaction and thus reaching the anaerobic stage. When most oxygen is removed, clostridial growth is discouraged and lactic acid fermentation encouraged.
- Chopping releases plant juices, stimulating the growth of lactic acid bacteria.
- Chopping may increase silage intake by improving quality of fermentation and by accelerating rate of passage of feed particles through the rumen.

– Chopped silage reduces the rumination and may decrease milk fat content. Thus, 10-15 percent of the silage material should be above 25 mm in length in order to maintain an effective fibre function.

## Quality of Silage

Quality of silage fermentation is influenced by several factors, including moisture, WSC content of raw materials, degree of compaction and effectiveness of final sealing. Stage of maturity influences a forage's nutritive value and thus quality of silage. Leguminous forages at budding stage have optimal energy, protein and carotene contents, but DM yield and nutritive value decrease with maturity (Table 1).

*Table 1. Yield of dry matter and nutrients in legumes and grasses by growth stage*

| Growth stage | Yield (ton/ha) | | | |
|---|---|---|---|---|
| | Fresh yield | DM | Feed unit | Digestible CP |
| *Legumes* | | | | |
| Pre-bud | 13.0 | 2.7 | 2.3 | 0.44 |
| Mid-bud | 16.5 | 4.2 | 3.8 | 0.58 |
| Flowering | 14.8 | 4.2 | 3.7 | 0.54 |
| Full bloom | 8.4 | 3.5 | 1.9 | 0.19 |
| Green seed pod | 7.1 | 2.8 | 1.2 | 0.13 |
| Milky ripe and dough seed | 6.8 | 2.8 | 0.5 | 0.10 |
| *Grasses* | | | | |
| Heading | 17.2 | 5.0 | 4.6 | 0.49 |
| Early bloom | 17.3 | 5.2 | 3.8 | 0.40 |
| Full bloom | 9.1 | 3.8 | 2.3 | 0.19 |
| Green seed pod | 8.4 | 3.0 | 1.2 | 0.08 |
| Milky ripe | 8.0 | 3.0 | 1.1 | 0.06 |
| Dough seed | 7.8 | 2.6 | 0.7 | 0.05 |

Grass at heading stage is highest in caloric value and protein content. When leguminous forages reach late flowering stage or grasses reach seed stage, the energy value has decreased by 70-75 percent, digestible CP by 67-83 percent, and carotene content by 75-84 percent.

Therefore, it is important to harvest forage crops at their optimal stages. Legumes should be harvested at budding stage, grasses at heading stage, and whole maize plant at late milky to early dough seed stage. Grass is highest in protein at pre-heading, and highest energetically at dough seed stage. Many grasses can not head again after the first cutting, and the re-growth may be collected 4 to 5 weeks after the first harvest. The appropriate stages at which plants may be harvested are indicated in Table 2.

*Table 2. Appropriate harvest stages for grasses and forage crops*

| Grass and forage crops | Stage of growth | Moisture (%) |
|---|---|---|
| Alfalfa | Late bud to 1/10 bloom | 70-80 |
| Red clover | Late bud to early bloom | 75-82 |
| Orchard grass | Pre-head to heading | |
| Awnless bromegrass | Pre-head to heading | 75 |
| Timothy | Pre-head to heading | |
| Sudan grass | About 90 cm in height | 80 |
| Mixed grasses | Pre-head to early heading | |
| Mixed legume and grass | Depending on grass | |
| Grain forage | Pre-head to early heading | |
| Whole-crop maize | Dough seed | 65-70 |
| Maize stover | Soonest after maize harvest | 50-60 |
| Whole-crop sorghum | Early to mid-dough seed | 70 |
| Sorghum stover | Soonest after sorghum harvest | 60-70 |
| Oat | Pre-head to early heading | 82 |
| Oat | Milky ripe | 78 |
| Oat | Early dough seed | 70 |
| Barley | Late bud to early dough seed | 82-70 |
| Rye | Late bud to early dough seed | 80-75 |

Because straw and stover are harvested after grain collection, their nutritional value is generally low. It has been recently shown that yield and grain quality do not change when harvest is shifted to an earlier stage (by 7-10 days). However, this shift may be beneficial in terms of improved nutritional value of straw and stover.

## Silage Additives

The purpose of using additives is to ensure silage quality by encouraging lactic acid fermentation, by inhibiting undesirable microbes or by improving its nutritional value. Common silage additives include bacterial cultures, acids, inhibitors of aerobic damage, and nutrients.

### *Bacterial Cultures*

A dominant lactic acid fermentation is the key to making good silage. Lactic acid bacteria are normally present on harvested crops together with clostridial bacteria in a ratio of about 10 to 1. Considerable nutrient loss usually occurs at initial stages of ensiling when oxygen is still present. Addition of lactic acid bacterial cultures during filling-up increases their population rapidly, encouraging lactic acid fermentation and pH reduction to a level that inhibits clostridial development. Different strains of lactic acid bacteria look similar under the microscope, but their biological activities are very different.

Only those acid-tolerant strains that possess a homo-fermentative pathway, producing the maximum amount of lactic acid from hexose sugars readily available, and a growth temperature range extending to 50°C, should be used as a silage additive. The environment under which lactic acid bacteria multiply favourably is also important. Lactic acid bacteria are anaerobic, and hence air should be removed and the silo should be kept airtight. These micro-organisms ferment soluble sugars to a mixture of acids, but predominantly lactic acid. Plant materials should contain at least 2 percent WSC, otherwise soluble sugars should be added.

Starch may also be added along with amylase in order to provide lactic acid bacteria with soluble sugars. Cereal straws and stovers are high in lignocellulose. A mixture of inoculum or enzymes, or both, containing cellulase and xylanase is often used. Enzymes degrade lignocellulose, liberating soluble sugars for bacterial use. A number of commercial preparations are available from foreign companies, and some have been registered by government authorities and can be sold in China. During the Ninth Five-Year Plan, Chinese scientists successful produced a special additive for ensiling fresh cereal straws and stovers.

#### *Acids*

The original proposal to use acids as silage additives dates back to 1885. In the late 1920s, Virtanen from Finland, adopted this approach and recommended the rapid acidification of the crop with mineral acids to a pH of about 3.5, which was originally thought to inhibit microbial and plant enzyme activity. This AIV process was widely used in Scandinavia until quite recently. Due to the difficulties in handling corrosive acids, organic acids were later used as silage additives. When acids are added, plant materials sink quickly and are easy to consolidate. Acidity may arrest plant respiration and reduce heat production and nutrient loss.

Rapid acidification may also inhibit clostridia. However, addition of acids increases effluent and can be potentially toxic to animals. Furthermore, acids are corrosive to people, animals and machinery. Reduction of moisture content may minimise effluent, and addition of calcium carbonate can be used to adjust silage acidity. Appropriate concentrations of different acids as silage additives are recommended as follows:

- sulphuric and hydrochloric acids: 50-80 litre of dilute acid (concentrated acid diluted with water at 1:5 v/v) per tonne
- formic acid: about 3 kg/ton
- acetic acid: 5-20 kg/ton
- propionic acid: 1 litre/$m^2$ of surface to prevent mould development.

#### *Inhibitors of Aerobic Deterioration*

The most common inhibitors of aerobic deterioration are sodium nitrate, sodium nitrite, sodium formate and formaldehyde. These chemicals do not contribute to the improvement of fermentation, but are effective in preventing silage deterioration. Some plant parts, such as larch leaves, contain natural bactericides that may safely function as antiseptics. Formaldehyde is a well-known sterilising agent and is commercially available as formalin, which contains 40 percent of the gas in aqueous solution. Scientists are interested in the additives because of their bacteriostatic properties, and because of their known properties of protecting plant proteins from rumen microbial degradation. Adding formaldehyde at 3-15 kg/ton may result in a well-preserved silage.

#### *Nutrients*

Nutrient additives are defined as substances which, when added to ensilage materials, contribute significantly to the nutritional value of the silage. Most of nutrient additives can also favour lactic acid fermentation. A number of materials are considered to be in this category.

*Nitrogenous compounds*

Certain crops, such as maize and most cereal straw, are nutritionally deficient in nitrogen, and when fed to ruminants as silage require supplementing with a protein supplement. An alternative approach is to improve the CP content of the silage by adding urea during ensiling. When applied to maize, urea produces silages with higher pH values and fermentation acid contents than in untreated silages. Urea addition has also a marked effect on nitrogenous components of the silages, resulting in higher CP, true protein, free amino acids and ammonia.

Addition of urea to cereal straw and stover may also have an ammoniation effect, which is associated with higher CP and lower fibre content. However, attention should be paid to the rapid release of ammonia from urea in the rumen. High concentrations of ammonia in the rumen may cause ammonia poisoning. One solution is to add the urea or other nitrogenous compounds together with soluble sugar sources, such as molasses and starchy grains. Urea may be added at 0.5 percent of fresh materials. When whole maize plant was added with urea at 0.5 percent, the CP of the resulting silage increased to 12.9 from 8.7 percent in the untreated silage. Urea may also be added prior to feeding the animals.

*Carbohydrate-rich materials*

Carbohydrate-rich materials are added to silage crops in order to increase the supply of available energy for the growth of lactic acid bacteria, and are of particular importance in crops such as legumes, which are deficient in soluble carbohydrate content. Materials that have been used for this purpose include molasses and cereals. Molasses is a by-product of the sugar industry, and has a DM content of 70-75 percent and a soluble carbohydrate content of about 65 percent of DM.

In order to obtain maximum benefit, it should be used at 4 percent (w/w) for grass silage and 6 percent (w/w) for legume silage. Cereals have been used as additives in an attempt to improve both the fermentation quality and the nutritional value of silages. Cereals contain 50-55 percent of starch that can be hydrolysed to soluble sugars and then utilised by lactic acid bacteria. If amylase or amylase-rich materials, such as malt, are added with the cereal to ensiled crops, they will be more effective as fermentable carbohydrate sources.

#### *Minerals*

In addition to being nutritionally deficient in N, most straw and stover is a poor source of Ca and many micro-elements. Limestone is sometimes added to silages as Ca supplement and to alleviate silage acidity. Calcium carbonate may be added at 0.45-0.50 percent to obtain maximum benefit. Common salt has a high osmotic pressure to which clostridia are sensitive, but lactic acid bacteria are not. Addition of salt may increase lactate content, decrease acetate and butyrate, resulting in silage with good quality and palatability.

Other minerals may be used in ensiled materials. Examples of mineral sources include copper sulphate (2.5 ppm), manganese sulphate (5 ppm), zinc sulphate (2 ppm), cobalt chloride (1 ppm) and potassium iodide (0.1 ppm). Finally, it has to be pointed out that recommendations on the use of silage additives must be based not only on the results of scientific research, but also on sound economic return.

### Silage Quality Evaluation

The nutritive value and the quality of silage should be accurately evaluated. Working on behalf of the Bureau of Animal Production and Health (BAPH), MOA, researchers at Zhejiang University drafted methods of evaluating nutritive value and quality of silage, which have been tested in China since 1996. The pH value and certain simple subjective criteria such as colour, smell and texture are used to evaluate the quality of silage on-farm.

#### *pH Value*

The pH is the simplest and quickest way of evaluating silage quality, and may be determined on-farm using wide-range pH test papers such as bromophenol blue (range 2.8-4.4), bromocresol green (range 4.2-5.6) and methyl red (range 5.4-7.0). The classification of silage based on pH value is:

| | |
|---|---|
| pH below 4.0 | excellent |
| pH between 4.1 and 4.3 | good |
| pH between 4.4 and 5.0 | average |
| pH above 5.0 | bad |

#### *Colour*

Good silage usually preserves well the original colour of the standing plant. When green raw material produces silage with green or yellow colour, it can be considered of good quality. Temperature is one of the important factors affecting silage colour. The lower the temperature during ensilage, the less colour change. Above 30°C, grass silage becomes dark yellow. Above 45 to 60°C, the colour becomes closer to brown.

Beyond 60°C, the colour darkens towards black due to caramelisation of sugars in the forage. However, silage quality can be misjudged by on a colour basis. For example, silage from red clover or Chinese milk vetch is often dark brown instead of light brown. Despite its excellent quality, it may be considered a failure due to colour. A more useful indicator is colour of the silage juices. It can generally be said that the lighter the colour of the juice, the greater the success.

#### Smell

Good silage usually has a mild, slightly acidic and fruity smell, resembling that of cut bread and of tobacco (due to the lactic acid). A rancid and nauseous smell denotes the presence of butyric acid and signifies a failed silage. A musty smell is a sign of deficient compaction and presence of oxygen. A distinctive unpleasant smell, of sow's urine and faecal matter, is always indicative of marked protein degradation during ensilage.

#### Texture

Plant structures (stems and leaves) should be completely recognisable in the silage. A destroyed structure is a sign of severe putrefaction. A viscous, slimy appearance reveals the activity of pectolytic (sporulating) micro-organisms.

#### Taste

This test is more suitable for specialists. It is of little value to the farmer, whose basis is the palatability to farm animals.

#### Integrative Evaluation

It is obvious that silage quality can not be satisfactorily evaluated on any one of the above subjective indicators. The following methods of integrated evaluation have been on trial. Tables 3, 4 and 5 present indicators of integrated silage evaluation from maize stover, Chinese milk vetch or alfalfa, and sweet potato vines, respectively. From overall evaluation scores (based on the parameters given below in Tables 3 to 5), silage quality can be classified as follows:

| Class | Good | Satisfactory | Average | Bad |
|---|---|---|---|---|
| Scores | 100-75 | 75-51 | 50-26 | Below 25 |

*Table 3. Integrated evaluation of silage from maize stover using scores*

| *Score* | *pH* | *Moisture* | *Smell* | *Colour* | *Texture* |
|---|---|---|---|---|---|
| | *25* | *20* | *25* | *20* | *10* |
| | 3.4(25), | 70%(20), | Pleasant | Light | Loose and |
| | 3.5(23), | 71%(19), | and | yellow | |
| Good | 3.6(21), | 72%(18), | sweet, acidic | | soft, non- |
| | 3.7(20), | | | | |
| | 3.8(18) | 74%(16), | (18-25) | (14-20) | viscous |
| | | 75%(14) | | | (8-10) |
| | 3.9(17), | 76%(13), | | | |
| | 4.0(14), | 77%(12), | Light acidic | Brown | (Medium) |
| | | | | yellow | |
| Satis-factory | 4.1(10) | 78%(11), | | | |
| | | 79%(10), | | | |
| | | 80%(8) | (9-17) | (8-13) | (4-7) |
| | 4.2(8), | 81%(7), | Irritative, | | |
| | 4.3(7) | 82%(6), | alcohol, | (Medium) | Slightly |
| Average | 4.4(5), | 83%(5), | acidic | | viscous |
| | 4.5(4), | 84%(3), | (1-8) | (1-7) | (1-3) |
| | | 4.6(3), 4.7(1) | 85%(1) | | |
| | | | Rancid | Dark | Putrefactive |
| | | | and | brown | & |
| Bad | Above | Above | musty smell | | aggluti- |
| | 4.8(0) | 86%(0) | | | native |
| | | | (0) | (0) | (0) |

*Table 4. Integrated evaluation of silage from sweet potato vines using scores*

| *Score* | *pH* | *Moisture* | *Smell* | *Colour* | *Texture* |
|---|---|---|---|---|---|
| | 25 | 20 | 25 | 20 | 10 |
| | 3.4(25), | 70%(20), | Pleasant | Brown | Loose |
| | 3.5(23), | 71%(19), | and | | and |

*Contd. . .*

| Good | 3.6(21), | 72%(18), | sweet acidic | | soft, |
|---|---|---|---|---|---|
| | 3.7(20), | 73%(17), | | | non- |
| | 3.8(18) 75%(14) | 74%(16), | (18-25) | (14-20) (8-10) | viscous |
| | 3.9(17), | 76%(13), | Light | (Medium) | (Medium) |
| | 4.0(14), | 77%(12), | acidic | | |
| Satisfactory | 4.1(10) | 78%(11), 79%(10), | | | |
| | | 80%(8) | (9-17) | (8-13) | (4-7) |
| | 4.2(8), | 81%(7), | Irritative, | (Medium) | Slightly |
| | 4.3(7), | 82%(6), | alcohol, | | |
| Average | 4.4(5), 4.5(4), | 83%(5), 84%(3), | acidic | | viscous |
| | 4.6(3), 4.7(1) | 85%(1) | (1-8) | (1-7) | (1-3) |
| | | | Rancid and | Dark brown | Viscous and |
| Bad | Above 4.8(0) | Above 86%(0) | musty smell | | agglutinative |
| | | | (0) | (0) | (0) |

*Table 5. Integrated evaluation of silage from Chinese milk vetch or alfalfa*

| *Score* | *pH* | *Moisture* | *Smell* | *Colour* | *Texture* |
|---|---|---|---|---|---|
| | 25 | 20 | 25 | 20 | 10 |
| | 3.6(25), | 70%(20), | Pleasant | Light | Loose and |
| | 3.7(23), | 71%(19), | and | yellow | |
| Good | 3.8(21), | 72%(18), | acidic | | soft, non- |
| | 3.9(20), | 73%(17), | | | |
| | 4.0(18) | 74%(16), 75%(14) | (18-25) | (14-20) | viscous (8-10) |
| | 4.1(17), | 76%(13), | Sour and | Golden | (Medium) |
| | 4.2(14), | 77%(12), | | | |
| Satisfactory | 4.3(10) | 78%(11), 79%(10), | alcoholic | yellow | |
| | | 80%(8) | (9-17) | (8-13) | (4-7) |
| | 4.4(8), | 81%(7), | Irritative | Light | Slightly |
| | 4.5(7), | 82%(6), | acidic | yellow | |

*Contd. . .*

| | | | | | |
|---|---|---|---|---|---|
| Average | 4.6(6), 4.7(5), 4.8(3), 4.9(1) | 83%(5), 84%(3), 85%(1) | (1-8) | brown (1-7) | viscous (1-3) |
| Bad | Above 5.0(0) | Above 86%(0) | Rancid and musty (0) | Dark brown (0) | Putre-factive & aggluti-native (0) |

***Chemical Methods***

Evaluation of silage in the laboratory is mainly based on chemical analysis. It includes pH determination and assay of organic acids, which are the main fermentation metabolites in silage. Fermentation characteristics may be estimated based on total acids or on proportions of individual acids. Free ammonia assay, and better still the ratio of ammonia nitrogen to total nitrogen, is the most valid criterion for protein degradation. Before discussing the value of such analyses, it is necessary to emphasise the necessity for the sample to be truly representative.

Samples must be taken with a probe from several places at different layers in the silos. The sampling points should be more than 30 cm from the edge of the silo to prevent misleading results. Samples should preferably be sent to the laboratory in sealed plastic bags or glass bottles to avoid aerobic deterioration (secondary fermentation). Ammonia nitrogen and pH should be determined as soon as possible.

*pH value*

The pH of silage should be measured in the laboratory using a precise pH meter. For fresh grass silage and maize silage at all DM contents, the pH is both the simplest and quickest method of evaluation. Research has shown that there is a very close relationship between pH value, fermentation quality and DM during ensiling.

### *Evaluation of Silage from Ammonia N*

The ratio of ammonia-N to total N ($NH_3$-N/TN) in silage provides information on the stage of protein degradation and it undeniably constitutes a test of the state of conservation of the ensiled proteins. The higher the ratio, the more protein has been degraded, and the worse the quality. In the system proposed, a maximum of 50 points is given for a ratio lower than 5 percent, and points are deducted for a ratio higher than 35 percent. Table 6 presents the point scale for the evaluation of silage from ammonia N.

*Table 6. Scale for silage evaluation from the ratio of ammonia-N to total N*

| Silage quality/ | % $NH_3$-N TN | Points | Silage quality | % $NH_3$-N /TN | Points |
|---|---|---|---|---|---|
| Very good | <5 | 50 | Average | 15.1-16.0 | 22 |
| Good | 5.1-6.0 | 48 | | 16.1-17.0 | 19 |
| | 6.1-7.0 | 46 | | 17.1-18.0 | 16 |
| | 7.1-8.0 | 44 | | 18.1-19.0 | 13 |
| | 8.1-9.0 | 42 | | 19.1-20.0 | 10 |
| | 9.1-10.0 | 40 | Bad | 20.1-22.0 | 8 |
| Satisfactory | 10.1-11.0 | 37 | | 22.1-26.0 | 5 |
| | 11.1-12.0 | 34 | | 26.1-30.0 | 2 |
| | 12.1-13.0 | 31 | Very bad | 30.1-35.0 | 0 |
| | 13.1-14.0 | 28 | | 35.1-40.0 | -5 |
| | 14.1-15.0 | 25 | | >40.1 | -10 |

### *Evaluation Using Various Organic Acids*

The presence of lactic acid and various volatile fatty acids, especially acetic, propionic and butyric acids, is a reflection of the fermentation that has occurred. It is appropriate therefore to take into account, when judging silage success, both the type of acids and the amount present. The higher the proportion of lactic acid, the better the quality.

*Table 7. Key for the evaluation of silage according to organic acids*

| Percent | Points | | | Percent | Points | | |
|---|---|---|---|---|---|---|---|
| | Lactic | Acetic | Butyric | | Lactic | Acetic | Butyric |
| 0.0-0.1 | 0 | 25 | 50 | 28.1-3 0.0 | 5 | 20 | 10 |
| 0.2-0.5 | 0 | 25 | 48 | 30.1-3 2.0 | 6 | 19 | 9 |
| 0.6-1.0 | 0 | 25 | 45 | 32.1-3 4.0 | 7 | 18 | 8 |
| 1.1-1.6 | 0 | 25 | 43 | 34.1-3 6.0 | 8 | 17 | 7 |
| 1.7-2.0 | 0 | 25 | 40 | 36.1-3 8.0 | 9 | 16 | 6 |
| 2.1-3.0 | 0 | 25 | 38 | 38.1-4 0.0 | 10 | 15 | 5 |
| 3.1-4.0 | 0 | 25 | 37 | 40.1-4 2.0 | 11 | 14 | 4 |
| 4.1-5.0 | 0 | 25 | 35 | 42.1-4 4.0 | 12 | 13 | 3 |
| 5.1-6.0 | 0 | 25 | 34 | 44.1-4 6.0 | 13 | 12 | 2 |
| 6.1-7.0 | 0 | 25 | 33 | 46.1-4 8.0 | 14 | 11 | 1 |
| 7.1-8.0 | 0 | 25 | 32 | 48.1-5 0.0 | 15 | 10 | 0 |
| 8.1-9.0 | 0 | 25 | 31 | 50.1-5 2.0 | 16 | 9 | -1 |
| 9.1-1 0.0 | 0 | 25 | 30 | 52.1-5 4.0 | 17 | 8 | -2 |
| 10.1-1 2.0 | 0 | 25 | 28 | 54.1-5 6.0 | 18 | 7 | -3 |
| 12.1-1 4.0 | 0 | 25 | 25 | 56.1-5 8.0 | 19 | 6 | -4 |
| 14.1-1 6.0 | 0 | 25 | 24 | 58.1-6 0.0 | 20 | 5 | -15 |
| 16.1-1 8.0 | 0 | 25 | 22 | 60.1-6 2.0 | 21 | 0 | -10 |
| 18.1-2 0.0 | 0 | 25 | 20 | 62.1-6 4.0 | 22 | 0 | -10 |
| 20.1-2 2.0 | 1 | 24 | 18 | 64.1-6 6.0 | 23 | 0 | -10 |
| 22.1-2 4.0 | 2 | 23 | 16 | 66.1-6 8.0 | 24 | 0 | -10 |
| 24.1-2 6.0 | 3 | 22 | 14 | 68.1-7 0.0 | 25 | 0 | -10 |
| 26.1-2 8.0 | 4 | 21 | 12 | >70 | 25 | 0 | -10 |

The evaluation system is basically that of Flieg from 1938 with a maximum of 100 points given. Full points are given for lactic acid (25) above 68 percent, for acetic acid (25) below 20 percent and for butyric acid (50) below 0.1

percent. Individual acids are scored independently, and the total is the sum of the three acids. Table 7 presents the key for silage evaluation according to Flieg (0 - 20 = failure; 21 - 40 = poor; 41 - 60 = satisfactory; 61 - 80 = good; 81 - 100 = very good.

### *Silage Evaluation from Ammonia-N and Organic Acids*

In order to integrate the information from both protein and carbohydrate in the silage, an evaluation system based on the points from ammonia-N and organic acids is proposed. There is a maximum of 100 points in the scale, 50 points for proteins and 50 points for carbohydrates. The following table is used for evaluation of silage quality from protein and carbohydrate:

| Total points | 0-20 | 20-40 | 40-60 | 60-80 | 80-100 |
|---|---|---|---|---|---|
| Overall quality | Very bad | Bad | Average | Good | Very good |

### *Feeding Silage*

Silage is suitable for feeding some 6-7 weeks after ensiling. The basic principle for silage use is to minimise the area exposed to air, with as little churning as possible. Silage should be taken from one side of the trench, and the exposed surface should be immediately covered to minimise re-entry of air and avoid secondary fermentations. Material taken out from the silo should be fed as soon as possible, since it rapidly deteriorates upon contact with air. Silage left in the trough must be cleaned away in time to prevent putrefaction.

Some animals may not like silage when offered for the first time. In these cases, some adaptive measures should be taken. For example, silage can be placed in the bottom of the trough, and covered with concentrates. When animals are adapted to silage, the offer can be increased. Silage can be fed to all kinds of animals:

replacement cattle, fattening cattle, dairy cows, sheep, goats and even pigs. The amount of silage offered depends on the animal and its age, as well as on the type and quality of silage. Taking cattle as an example, the recommended amount to be fed daily is:

| | |
|---|---|
| Grass silage | 4 kg per 100 kg liveweight, |
| Legume silage | 3 kg per 100 kg liveweight, |
| Maize stover silage | 4 kg per 100 kg liveweight, |
| Starch-rich forage | 5 kg per 100 kg liveweight, |

Therefore a cow with liveweight of 500 kg may be daily fed 20 kg maize stover silage. Sometimes the milk from cows fed on silage has a taint. Because this smell is only transmitted through the air, the following points should be observed to prevent it:

1. *Handling* Silage should never remain in the cowshed, and it should be offered in amounts exactly as required.
2. *Feeding method* Silage should not be given before milking. The effect on milk taste is more marked when the silage is fed 2 hours before milking, and least when given 6 hours after.
3. *Hygiene and cleanliness* Both floor and operators should be clean and the shed well ventilated.
4. *Milk and milking* Equipment should be kept clean, and milk cooled as soon as possible.

It is sometimes believed that ensiled forage has an adverse effect on the general health of the animal, especially on the skeleton of young animals. This criticism has no scientific evidence. It is certain that musty silage adversely affects animals, but such material should never be fed in the first place. It must be pointed out that silage is a good feed, but it is certainly not a "complete feed." For example, maize silage contains insufficient Ca and P, and should be supplemented. P content in ensiled alfalfa is sufficient for heifer growth and Ca more than enough.

Animals fed on silage should be properly supplemented with the nutrients that are insufficient for the animals' requirements. It must also be emphasised that the nutritional value of silage depends on the quality of the original forage, harvest time, conservation method, etc. Nutritional values of different silages therefore vary substantially. When it is impossible to analyse silage at the laboratory, one should refer to feed manuals and roughly assess the nutritional value of the silage to be fed, in order to formulate a ration that matches the animal's requirement.

# 6

# Ammoniation of Crop Residues

In the 1950s, a Danish patent was issued on ammonia treatment technology. Chinese farmers have known that human urine can be used to treat straw for cattle feed. It is pity that no research has been done on this. From the 1980s, China started to adjust its animal production structure with a new policy of a grain saving strategy. Fundamental research and experiments were started. MOA began to popularise straw ammoniation in 1987.

In 1989, straw ammoniation became one of ten key extension techniques of MOA, and by 1993 it had been popularised throughout the country, with 11.7 million tonne of straw treated, the largest in the world. FAO and UNDP had several successful projects on straw utilisation for feeding animals. At FAO's suggestion, international conferences on *Increasing Animal Production with Local Resources* were organised in China on three occasions. These conferences provided the opportunity to exchange experiences among different countries.

### Effects of Straw Ammoniation

The main component of straw is fibre, including cellulose and hemicellulose that can be digested by ruminants. Some cellulose and hemicellulose are bound to lignin and resistant to microbial attack. The role of ammoniation is to destroy this link, so these fractions are available to the animal. Ammoniation usually increases digestibility by 20 percent and CP content up to 1-2 times. It can also improve palatability and consumption rate. The total nutritional value can be doubled, reaching 0.4-0.5 feed units for each kilogram of ammoniated straw. In addition, ammoniation reduces mould development, destroys weed seeds (e.g. wild oat, false sorghum, etc.), parasite eggs and bacteria. The sources of ammonia to treat straw include anhydrous ammonia, urea, ammonium bicarbonate and aqueous ammonia.

### *Anhydrous Ammonia*

Anhydrous ammonia means "ammonia without water." Its formula is $NH_3$, and its N content is 28.3 percent. The normal dosage is 3 percent by weight of the straw DM. It is the most economical source of ammonia. The boiling point of anhydrous ammonia is -33.3°C, its vapour density is 0.59 and its liquid density 0.62. Gas pressure is 1.1 kg/cm$^2$ at -17.8°C and 13.9 kg/cm$^2$ 38°C. At normal temperature and pressure, anhydrous ammonia is a gas.

Expensive pressure containers are required not only to keep it as a liquid, but also to transport and store it. Anhydrous ammonia is a potentially dangerous and toxic material, and stringent safety precautions need to be observed when using it. Its natural ignition temperature is 651°C. If the ammonia content in the air reaches 20 percent, an explosion from self-ignition could occur. Attention should be paid to possible ammonia explosions, even though it seldom happens.

### *Urea*

The N content of urea is 46.7 percent. Its formula is

$CO(NH_2)_2$. It is decomposed into ammonia and $CO_2$ by ureases at ambient temperature. The chemical reaction is:

$$CO(NH_2)_2 + H_2O \xrightarrow{\text{Ureases enzymes. Ambient temperature}} 2NH_3 / +CO_2$$

Urea dosage needed to treat straw may vary a lot. The recommended dosage is 4-5 percent urea on DM basis, taking into consideration the effect of ammoniation and costs. Urea can be transported conveniently at normal temperature and pressure. It is harmless to humans. Treating straw with urea does not need complex equipment and the sealing conditions are not as strict as with anhydrous ammonia. It is known that farmers in Bangladesh ammoniate straw in bamboo baskets lined and covered with leaves of a kind of banana.

From the *Yellow Cattle Magazine*, it is known that, in Anhui province, technicians use urea as a source of ammonia to treat straw without cover and get good results, which is beneficial for extension of straw ammoniation to rural areas. At present, urea is a widely used source of ammonia in China. Urea is not as effective as anhydrous ammonia for straw treatment, but it is better than ammonium bicarbonate.

### Ammonium Bicarbonate

The nitrogen content of ammonium bicarbonate is 15-17 percent; its formula is $NH_4HCO_3$. It can be decomposed into $NH_3$, $CO_2$ and $H_2O$ at a suitable temperature. The chemical reaction is:

$$NH_4HCO_3 \xrightarrow{\text{Heating}} NH_3 / +CO_2 + H_2O$$

The dosage of ammonium bicarbonate, estimated by its N content, is 14-19 percent of straw DM. But, according to the experiments carried out in Zhejiang and Shanxi Agricultural Universities, using 8-12 percent of ammonium bicarbonate can give the same result as with 14-19 percent. Ammonium bicarbonate is a major product of the fertiliser industry and it is readily available at low price. It has a retail price of ¥ 300/ton and it is easy to use. Since ammonium bicarbonate is an intermediate product of urea breakdown, theoretically, in the right concentration, its effect should be similar to urea.

Reports from Zhejiang Agricultural University indicate that in the humid south, straw ammoniated with urea showed more mould spots than with ammonium bicarbonate. It does not decompose completely at low temperature, thus in cold climates the effectiveness of treatment with ammonium bicarbonate is not good. When treating with ammonium bicarbonate in an oven, one day is enough, since the temperature reaches 90 °C and it decomposes completely.

#### *Aqueous Ammonia*

Aqueous ammonia is a solution of ammonia in water. The concentration is quite variable, but the usual value is 20 percent. At this concentration, the normal dosage is 12 percent by weight of straw DM. It is only adapted to areas near to fertiliser factories because its low N content makes transport expensive. Besides the above sources of ammonia, human and animal urine also can be used to treat straw. However, collection difficulties limit practical applications.

### Methods for Ammonia Treatment

At present, the methods for popular ammonia treatment of straw in China include stack, silo or bunker, and oven methods. Each uses a different ammonia source.

***Stack Method***

The procedure for the stack method is as follows. First, an area is selected with an elevated, dry and even surface. This area is covered with non-toxic polyethylene sheet and the baled or loose straw is stacked on it. It is better to bale straw or to chop it into pieces. Especially for hard and thick maize stover, chopping before treatment facilitates feeding, saves plastic and reduces the danger of puncturing the plastic. The moisture content of straw should be adjusted to 20 percent or more during stacking (anhydrous ammonia requires a low-moisture straw compared with urea and ammonium bicarbonate).

The high moisture has a positive effect on straw treatment. However, after treatment, it is difficult to ventilate the straw and it can be easily attacked by mould. A wooden bar, which will be pulled out when ammonia injection starts, is placed before stacking so that ammonia can be injected easily and conveniently. The stack is sealed with non-toxic polyethylene sheet and injected with 3 percent anhydrous ammonia by weight of straw DM. At the end, the hole left when the pipe is removed is sealed with good quality tape.

Operators working with ammonia must be trained and should strictly follow relevant regulations to ensure personal safety. Treatment time varies inversely with temperature. For example, more than four weeks are needed at a temperature of 5-15°C, but only one week at >30°C. Straw treatment with anhydrous ammonia is simple and efficient, but particularly suited for large-scale use. The method also had been widely applied in developed countries. It requires some expensive equipment.

For instance, the cost of an EQ144 ammonia truck tank carrying 5.1 tonne is ¥ 16 000; a 200-kg ammonia bottle costs ¥ 2 500; a 400-kg bottle costs ¥ 4 500; and an 8.82 kW four-wheel tractor costs ¥ 8 000. Furthermore, anhydrous ammonia is not well suited for private farmers because of its dangers. In order to explore the best supply system of ammonia for farmers, ammoniation stations have been built in Hebei Xingtai region, Boxiang county; Baoding region, Dingxing county; and Henan Zhoukou region, Fugou county. These stations have accumulated experiences during the past few years to share with other regions.

*Selection of plastic film*

The basic requirements for plastic film are that it is non-toxic, durable and suitable for sealing. The plastic often used is polyethylene. Thickness, width and colour are determined by practical situations. Thick film (about 0.12 mm) is used for maize stover; thin film (less than 0.12 mm) for wheat straw. Width of film is determined by the size of stack and market availability. If used in the open air, black colour should be preferred, because it is durable and absorbs solar energy, which heats the stack and shortens treatment time. If used indoors, film colour has no obvious influence on treatment. The amount of film required can be calculated by the size of stack.

| | |
|---|---|
| Size of bottom sheet: | Length = Length of stack + (0.5-0.7) m (overlap) |
| | Width = Width of stack + (0.5-0.7) m |
| Size of covering sheet: | Length = Length of stack + height x 2 + (0.5-0.7) m |
| | Width = Width of stack + height x 2 + (0.5-0.7) m |

The ammoniation station in Fugou County made a covering sheet that was especially suitable for a 500-kg wheat straw stack. Covering the stack greatly improves treatment efficiency.

*Stack density measurement*

Weighing the stack is a basic task for straw treatment. It is well known that it is important to inject the correct amount of ammonia: too little ammonia is ineffective; too much ammonia increases the cost and has no further effect on treatment. The exact weight of straw must be known so that the correct amount of ammonia can be applied. But weighing is difficult under field conditions. A simple method is to first measure the average density of stack for various straws, then to multiply it by its volume.

Stack density depends upon plant species, moisture content and particle size. Of course, density also varies with time. In order to get reliable data, it is necessary to measure many stacks. Density is expressed in $kg/m^3$. In old stacks, volume is measured first, then it is weighed. New stacks are weighed before stacking and volume measured after. It is very easy to calculate the volume of rectangular and cylinder stacks. The volume of stack with irregular shape may only be estimated. In 1989, 9 stacks of air-dried straw were measured in Boxiang county during the FAO project. The average density was 55 and 79 $kg/m^3$ for new and old wheat straw, and 79 and 99 $kg/m^3$ for new and old maize stover, both whole, respectively.

*Ammonia dosage measurement*

The precision of ammonia injection depends on not only the correct weight of straw but also the correct amount of ammonia. Currently there are two main methods for ammonia injection used in China. One is to inject ammonia into the stack directly from an ammonia truck tank filled in the factory (Figure 1), the other is to inject it from a bottle (Figure 2).

*Figure 1. Transfer of anhydrous ammonia from tank truck to stack of crop residues*

At present, domestically produced ammonia flow meters are not available in China, and the price of imported ammonia flow meters is very high.. The Non-conventional Feeds Institute of China Agricultural University has developed an ammonia flow meter. With a flow of 3-5 kg/min, the error is less than 3 percent. The price of the meter and its accessories is ¥ 500 and it is ready to go to the market.

In most situations, the amount of ammonia to be injected is estimated by the injecting time. This method is not very precise because both the pressure and the flow rate of the ammonia vary with temperature. Sometimes ammonia dosage is estimated by the weight of the ammonia bottle. This method is more precise for ammonia injection. In brief, there are still many problems to be solved in terms of measuring ammonia dosage.

In order to solve the measurement problems of ammonia dosage, F. Sundstol, FAO expert, proposed "one stack, one bottle." This approach could overcome the measurement difficulties, since a whole bottle would be used for a standard stack. At the same time, small bottles should be easily transported by small tractors with trailer, or by bicycle, an obvious advantage for rural areas with bad roads. In addition, farmers can buy ammonia bottles and store them for later use.

*Figure 2. Transfer of anhydrous ammonia from ammonia bottle to stack of crop residue*

For the small ammonia bottle system it is very important to select the appropriate size. If the ammonia dosage is 3 percent of the straw, the net weight of ammonia should be 15, 30 and 60 kg for 500, 1 000 and 2 000 kg stacks, respectively. A large stack should have a weight multiple of the small stack, so that entire bottles can be used. The "one stack, one bottle" system has been tried in Henan, Hebei and other regions.

It has been found that there are still some problems to resolve before it is generalised. Firstly, the small ammonia bottle is thin and long, inconvenient for transport by bicycle. Secondly, the popularisation of the "small ammonia bottle" implies more people involved in their transport, with increased accident risk. It is also imperative to strengthen training for safety operation.

#### *Silo or Bunker Method*

Ammoniating straw with urea in silos or bunkers has been widely used in China. This method has many advantages: silos or bunkers may be used for either ammoniating or ensiling year round; bunkers are easy to manage and avoid rodent damage to plastic films. Silos or bunkers constructed with cement are the best, since they save on plastic and minimise repairs. Once a silo or bunker is constructed, it can be used for several years. In addition, the bunker facilitates the estimation of straw weight. Animal type and quantity determine bunker size.

It should be known how much straw (air dried) can be placed per $m^3$ of bunker; how much ammoniated straw an animal requires per year; whether the bunker may also be used for ensiling; and how many treatments are required per year. Average weight of air-dried and chopped crop residue (wheat, rice and maize), measured in different regions, is about 150 kg per $m^3$. Straw intake varies with animal type, size and concentrate level. Generally speaking, daily straw intake is 2 to 3 percent of liveweight. For example, daily net straw intake is 4-6 kg for a 200 kg steer, equivalent to 1.5-2 tonne per year.

If the silo is used for straw, 1 $m^3$ will hold about 650 kg (25 percent DM). With these data, the size of a silo or bunker can be designed according to practical needs. There are many types of bunkers. They can be built on the surface, underground, or half-and-half. It is recommended to build rectangular bunkers that, by adding internal walls, can be divided into double or twin bunkers for sequential treatment (Figure 3). If a double bunker is used for silage, a second fermentation can be started in the second compartment while straw is being used from the other half. Double bunkers are very common in Henan Zhoukou region.

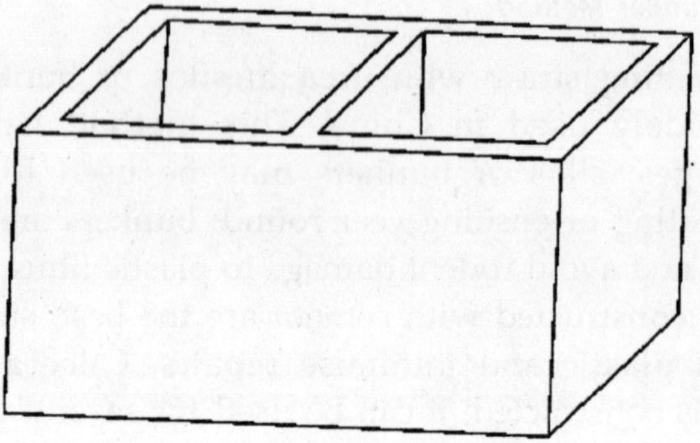

*Figure 3. A twin bunker*

For instance, a bunker with a volume of 2 $m^3$ requires 500 bricks, a bag of cement and a wagon of sand. Its total cost is ¥ 100. The cost of a double-bunker holding 4 $m^3$ is ¥ 200, affordable to farmers. One bunker of 2 $m^3$ can hold 300 kg of wheat straw. The two bunkers can be used in turns. One bunker of ammoniated straw is enough for two cattle for a month. If a bunker is used for silage, one bunker can hold more than 1 000 kg. The silo or bunker method has been well received by farmers.

*Operation*

First, straw is chopped to about 2 cm long. The general principle is that thick and hard residues, such as maize stover, should be cut shorter, while soft materials may be a little bit longer. Then urea (or ammonium bicarbonate) is added to water and stirred to completely dissolve it. Normally, 100 kg of dried straw needs 5 kg urea and 40-60 kg water. Next, the urea solution is sprayed repeatedly over the straw. Before loading the silo, straw can be spread in an open area to facilitate uniformity in spray application. While straw is added to the silo, each layer should be compacted till the bunker is full, and then it is covered with plastic film, held firmly in place by a layer of fine soil (Figure 4). Treatment time with urea is a little longer than with anhydrous ammonia.

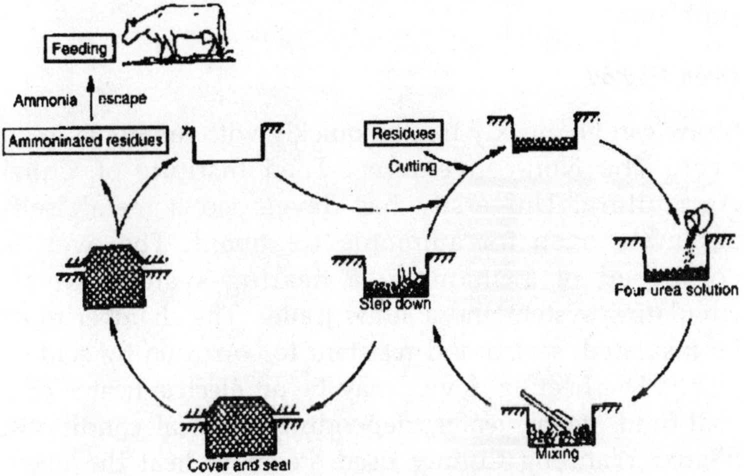


*Figure 4. Flow chart of the ammoniation of crop residues with urea*

When treating straw with urea, the speed of urea decomposition into ammonia should be taken into consideration. It depends on ambient temperature and amount of ureases present in the straw. Decomposition rate increases with temperature, so this method is well adapted to warm regions. In general, it is suggested to add some substance rich in ureases, such as soybean cake powder, to accelerate urea breakdown. Maize stover contains much more ureases than other cereal straws, so ureases are not required.

In China, maize stover is one of the three main crop residues, along with rice and wheat straw. The nutritive value of maize stover is higher than other straws, either before or after ammoniation. However, it was seldom used as feed because it is thick and hard, and not easy to be store and transport. Nowadays it should be preferentially used as feed because processing and treatment methods are available to overcome these problems.

### Oven Method

Straw can be quickly treated quickly with ammonia in an oven. The Non-conventional Feed Institute of China Agricultural University has developed a metal, self-assembly oven for ammonia treatment. The oven is composed of a chamber, a heating system, an air circulating system and a straw trailer. The chamber must be insulated, sealed and resistant to corrosion by acid or alkali. The heating device may be an electric heater or a coal-fired steam heater, depending on local conditions. Shanxi Wanrong County used steam to heat the oven, with good results. The straw trailer should be convenient for loading, unloading, transport and heating. A metal mesh trailer with steel wheels is the preferred option.

#### *Operation*

Ammonium bicarbonate (8-12 percent of straw DM) is first dissolved in water. Straw is placed on the trailer. The solution is uniformly sprayed over the straw, adjusting moisture content to about 45 percent. Once the trailer is full, it is moved into the chamber, the door is closed and heating started. If heating is by electricity, the heating tube needs to be set to control the chamber's temperature at about 95°C. After heating for 14-15 hours, it is turned off. The chamber should remain sealed for 5-6 hours more, and then the straw trailer is moved outside for ventilation.

Once ammonia residues are gone, the straw can be fed. The oven can shorten treatment time considerably since it only takes 24 hours to treat straw. In addition, oven treatment is not weather dependent, so ammoniation can be done all year round. However, the cost is relatively high, limiting its application. In recent years in Jilin Province, straw has been treated in a coal-heated chamber usually used for flue-curing tobacco. The investment for the oven was avoided and fuel expenses reduced, so treatment cost was greatly decreased.

Besides the above three widely used methods, there are still some other methods for straw treatment. For instance, ammoniation within plastic bags was readily accepted by farmers due to low initial cost. However, repeated purchases of plastic bags increased costs and restricted wider application. In some places, straw is treated in locally available containers. This latter method may be worth advocating.

## Influence of Factors on Ammoniation

There are a number of factors that influence the effectiveness of straw ammoniation, including dosage, moisture content, temperature and treatment time.

### *Ammonia Dosage*

Experiments relating ammonia dosage to digestibility were conducted by Sundstol et al. Results showed a significant improvement in in vitro digestibility by increasing ammonia dosage from 1 to 2.5 percent of straw DM, and a slight improvement from 2.5 to 4 percent. There were no further effects beyond 4 percent. In recent years, many similar experiments have been carried out with similar results. The economic optimum ammonia dosage probably lies between 2.5 and 3.5 percent of DM.

When treating straw with anhydrous ammonia, urea, ammonia bicarbonate or aqueous ammonia, the dosage should be estimated by their nitrogen content: 82.3, 46.7, 15 and 20 percent, respectively. As the conversion ratio of nitrogen to ammonia is 1.21, the dosage of other sources can be calculated by the following equation:

$$\text{Dosage (kg)} = \frac{\text{economic optimum for dosage of ammonia}}{\text{nitrogen content of ammonia source} \times 1.21}$$

Optimum dosages (per 100 kg of straw) are: 2.5-3.5 kg for anhydrous ammonia; 4.5-6.2 kg for urea; 13.8-19.3 kg (8-12 kg in practice) for ammonium bicarbonate; and 10.3-14.5 kg for aqueous ammonia (20% N). Normally, treating 100 kg air-dry straw requires either 3 kg of anhydrous ammonia, 4-5 kg of urea, 8-12 kg of ammonium bicarbonate or 11-12 kg of aqueous ammonia (20% N).

#### *Moisture Content*

Moisture content of straw is another important factor determining the effectiveness of treatment. Ammonia combines with water to form ammonium hydroxide $(NH_4)OH$. Regarding the optimum moisture content of straw, experts from China and abroad have different opinions. Sundstol *et al.* found that increasing moisture content of straw from 12 to 50 percent had a positive effect on *in vitro* organic matter digestibility (IVOMD) (Figure 5), regardless of temperature.

With extremely low moisture, ammonia treatment had no positive effect on enzyme solubility and straw IVOMD. Sundstol and Ekeern found that a 6-week treatment with 2 percent anhydrous ammonia of straw with 2.5, 5.0, 7.5 and 10.0 percent moisture, produced corresponding IVOMDs of 52.1, 58.5, 59.1 and 66.0 percent, respectively. Treating straw with 15-20 percent moisture with aqueous ammonia was better than with anhydrous ammonia. Aqueous ammonia improved the moisture content of straw. For example, using 2 percent aqueous ammonia to treat straw is equivalent to adding 6 percent water to straw.

Experiments proved that the straw IVOMD could be gradually improved by increasing moisture from 2.5 to 50 percent. Experiments conducted by Professor Ji Yilun of Shanxi Agricultural University indicated that the optimal moisture content of straw was about 45 percent when treated with urea or ammonia bicarbonate. High moisture contents would result in handling problems and greater risk of damage during storage.

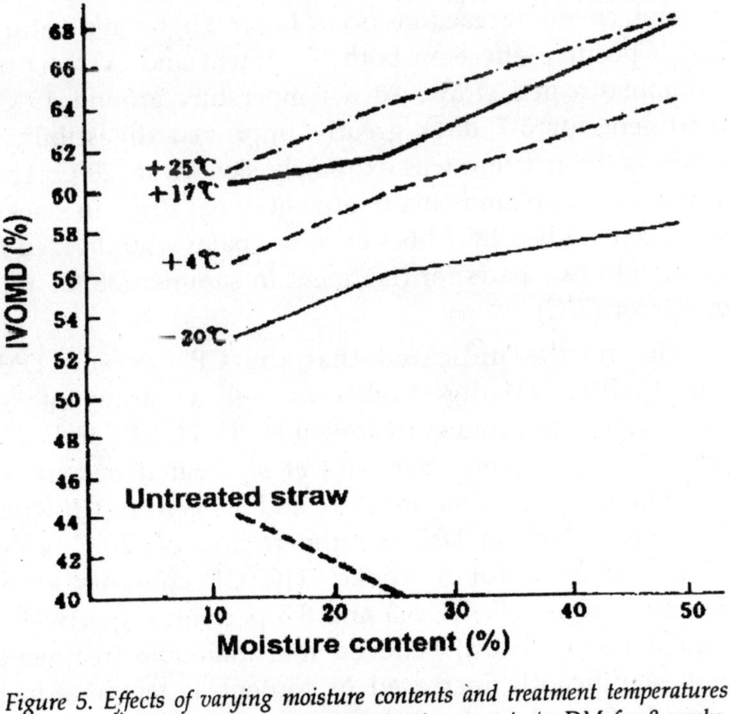


*Figure 5. Effects of varying moisture contents and treatment temperatures on straw IVOMD when treated with 3.4% ammonia in DM for 8 weeks Temperature and treatment time*

The positive effect of higher moisture has also been found in practice. Straw from both the top and the bottom of the stack has high digestibility and good intake due to its moisture content. Straw treated by the Animal Bureau of Zhoukou Region in Henan Province during the FAO project in 1989, with the proportions 100 kg straw: 100 kg water: 5 kg urea, gave good results despite the high moisture content. On the whole, higher moisture content of straw may improve digestibility. Moisture content can reach 50 percent or more, if straw can be transported and stored without becoming mouldy.

### *Effects of Temperature*

High temperature reduces treatment time, since, in general, chemical reactions occur faster. High temperature had a positive effect on both N content and IVOMD of ammonia-treated straw. At a temperature around 45°C, treatment for 3-7 days greatly improved digestibility, whereas the reaction was extremely slow near -20°C. The season effect on ammonia treatment of oat straw in stacks had been studied by Alibes *et al.* in Spain, who divided 6 tonne into two parts for treatment in summer (38°C) and in winter (7°C).

The results indicated that the CP content, DM digestibility, OM digestibility as well as straw intake were higher for summer treatment by 83, 12, 12.3 and 19.3 percent, respectively. Sundstol *et al.* treated oat straw containing 12.5 percent moisture and 3.4 percent CP with 3.4 percent $NH_3$ in DM at temperatures of -20°C, 4°C, 17°C and 24°C for 8 weeks. The CP contents after treatment were 6.5, 7.3, 8.3 and 8.5 percent, respectively. Sundstol *et al.* (1978) believed that ammonia treatment may significantly increased N content at temperatures over 25°C, but in longer treatment times increasing temperature from 40°C to 120°C had no positive effect on

IVOMD. The length of treatment time for different temperatures had been given by Sundstol *et al.*:

| *Temperature (°C)* | *Treatment time* |
|---|---|
| < 5 | >8 weeks |
| 5-15 | 4-8 weeks |
| 15-30 | 1-4 weeks |
| > 30 | < 1 week |
| > 90 | < 1 day |

A closed stack may be stored for a long time without becoming mouldy, and stacks should not be opened before it is necessary to use the treated feed. After opening a stack of ammonia-treated straw, it should be aerated for a time before feeding. In cold climates, increasing treatment time has positive effects on ammoniation, but CP content decreases slowly after opening. Beijing Dairy Institute treated rice straw with 3 percent hydrous ammonia and the CP content of treated straw reached 7.8 percent, but 3 months later was below 7 percent.

#### Pressure

High pressure is beneficial for ammonia treatment. Experiments conducted by Lie indicated that increasing the pressure from 1 to 5 kg/cm$^2$ had a positive effect on IVOMD. Sundstol and Owen had carried out experiments with similar results. It is known that if ammoniated straw is wafered or pressed into pellets, N-content and IVOMD can be further improved.

#### Quality of Ammoniated Straw

There are three methods to evaluate quality of ammoniated straw: sensory evaluation, chemical analysis and biological tests. The physical changes in straw can be visually observed as an easy, but imprecise, way to evaluate treated straw quality. By chemical analysis, the components of straw such as CF and CP can be measured, but by itself it can not give an estimate of overall nutritive value and animal intake. The scientific method of evaluation of straw quality is through biological tests. For example, digestibility *in sacco* (in the rumen), cellulase digestibility test, and especially digestibility *in vivo*, can estimate not only the digestibility but also the speed of digestion.

### *Sensory Evaluation*

Properly ammoniated straw is soft, brownish yellow or light brown, and with a light fragrance after excess ammonia has evaporated. If straw appears white or grey and is sticky or clumps, it means that it has been attacked by mould. This damaged straw should not be used as feed. Of course, this situation seldom occurs if treatment is correct. Mould normally results from high moisture content, defective sealing or delayed ventilation after opening. If, after ammoniation, straw colour is nearly the same as before treatment, it means that ammoniation did not go very well, but it can still be used as feed.

### *Chemical Analysis*

Chemical analysis is being widely used in China at present. This method evaluates the extent of straw quality improvement by analysing the main parameters before and after ammoniation. Lu Xilei, from the Beijing Dairy Institute, ammoniated straw by ensiling with anhydrous ammonia at 3 percent DM on the farm. Comparisons of the main parameters between ammoniated straw and Chinese wild-rye hay are shown in Table 1. After treatment CP content improved by 5.44, 3.98 and 5.02

percent for wheat straw, rice straw and maize stover, respectively. The corresponding figures for digestibility were 10.3, 24.0 and 20.0 percent, respectively. Digestibilities of ammoniated wheat straw and maize stover were nearly the same or higher than Chinese wild-rye hay.

*Table 1. Nutritional comparison between ammoniated straw and Chinese wild-rye hay*

| Feed | Nutritive factor (%) | | | |
|---|---|---|---|---|
| | DM | CP | CF | DM digestibility |
| Chinese wild-rye hay | 90 | 5.9 | 32.0 | 52.0 |
| Winter wheat straw | 90 | 2.2 | 41.0 | 39.7 |
| Ammoniated winter wheat straw | 70 | 7.6 | 39.0 | 50.0 |
| Rice straw | 93 | 3.9 | 33.1 | 24.0 |
| Ammoniated rice straw | 90 | 7.8 | 32.5 | 48.0 |
| Maize stover | 90 | 3.7 | 30.5 | 42.0 |
| Ammoniated maize stover | 90 | 8.7 | 30.5 | 60.0 |

#### Biological Tests

Biological tests, especially the nylon bag *in sacco* method for digestion rate, has been widely applied not only in science and research institutes but also in some production units. For example, in Henan Zhoukou region, the straw degradation rate was measured by the nylon bag technique in sheep during the FAO project. The results, presented in Table 2 and Figure 6, showed that the maximum rate of degradation (P-value) for ammoniated barley reached 77.1 from 52.1 percent in untreated straw. In the last five years, the Institute of Animal Production and Health in Hebei Province has measured digestibility rates for 9 feedstuffs by the nylon bag technique in sheep. These feeds include beans, cakes, pomace, dregs, animal by-products, leaf and straw. The nylon bag technique provides a foundation for systematic development of the feed resource.

*Table 2. Nylon bag (in sacco) dry matter degradation of wheat straw (%)*

| Sample | Incubation time (hours) | | | | | | a | b | c | RSD |
|---|---|---|---|---|---|---|---|---|---|---|
| | 0 | 8 | 24 | 48 | 72 | 96 | | | | |
| Ammoniated | 16.4 | 23.7 | 40.0 | 55.6 | 64.8 | 70.1 | 13.1 | 64.4 | 0.04 | 1.70 |
| Untreated | 12.2 | 17.0 | 32.7 | 44.1 | 48.1 | 50.8 | 4.9 | 47.3 | 0.02 | 0.99 |

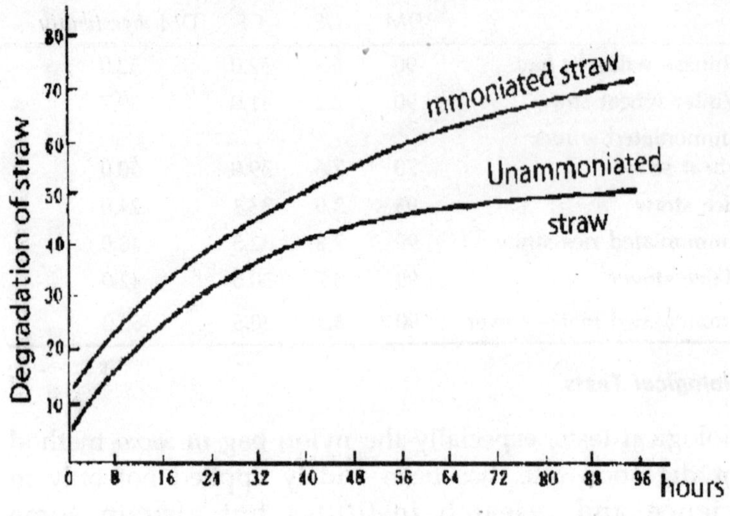


*Figure 6. Digestibility curve*

Straw degradation rate can be calculated by the equation:

$$p = a + b\,(1 - e^{-ct})$$

where:

p =the amount degraded at time (t)

a =soluble material which is immediately digestible,

b =the fraction which will be digested in the given time,

c =the rate constant for digestion of b,

a + b = the asymptote or the potential extent of digestion,

e =natural log

t =degradation time

The above parameters give an indication of feed quality. Considering the popularity of ammoniation, it is imperative to establish a standard suitable for estimating treated straw quality.

## Animal Experiments with Ammoniated Straw

### *Feeding Experiments with Beef Cattle*

A UNDP-funded project, *Beef cattle production system based on ammoniated straw,* was effected between 1991-1992 in Huaiyang and Shangshiu counties of Henan Province; Boxiang county in Hebei Province, and some other regions. At the same time, beef cattle feeding experiments were conducted.

Experiments were conducted in more than 100 counties with 259 animals. Cattle had *ad libitum* access to ammoniated straw and received a concentrate supplement according to body weight. Each animal needed 0.5 to 2.25 kg of concentrate feed (3/4 cotton seed cake + 1/4 maize grain). Results of feeding experiments are presented in Table 3, and the economic benefits in Table 4.

Results showed that average daily gain between 150 and 450 kg was 0.5 kg. The ratio of concentrate to weight gain was in the range of 0.83 to 3.5. This ratio was less than 2 for cattle under 350 kg. It took 498 days to go from 150 to 450 kg. Experiments with calves were also conducted. The results demonstrated that it took less than 2.5 years from birth to market weight (450 kg) under the current management regime. It took less than 2 years to bring cattle to 400 kg.

*Table 3. Feeding experiments with ammoniated straw in Huaiyang county*

| *Body weight (kg)* | *Concentrate (kg/day)* | *Straw intake (kg/day)* | *Weight gain (g/day)* | *Concentrate per kg gained* |
|---|---|---|---|---|
| 151-200 | 0.5 | 5.2 | 605 | 0.83 |
| 201-250 | 0.7 | 6.0 | 574 | 1.22 |
| 251-300 | 0.9 | 6.6 | 593 | 1.52 |
| 301-350 | 1.3 | 7.1 | 654 | 1.99 |
| 351-400 | 1.8 | 7.4 | 514 | 3.50 |
| 401-450 | 2.3 | 7.9 | 707 | 3.18 |

*Table 4. Economic analysis from studies in Huaiyang county*

| *Body weight (kg)* | *Weight gain (g/day)* | *Inputs (¥/day)* | *Output (¥/day)* | *Benefit (¥/day)* |
|---|---|---|---|---|
| 151-200 | 605 | 0.78 | 2.30 | 1.52 |
| 201-250 | 574 | 0.94 | 2.18 | 1.24 |
| 251-300 | 593 | 1.08 | 2.25 | 1.17 |
| 301-350 | 654 | 1.30 | 2.48 | 1.18 |
| 351-400 | 514 | 1.53 | 1.95 | 0.42 |
| 401-450 | 707 | 1.77 | 2.69 | 0.92 |

In Shangshui county the experiments were carried out on 85 private farmers, with 344 cattle, during 1992. Cattle had free access to ammoniated straw mixed with concentrate according to body weight. Each beast consumed 1.5 to 2 kg concentrate (3/4 cotton seed cake + 1/4 ground maize) per day. The results of feeding experiments are shown in Table 5, and the economic benefits in Table 6. Average daily gain was between 608 and 629 g from 150 to 450 kg when cattle consumed ammoniated straw supplemented with 1.5-2 kg concentrate per day. The ratio of concentrate to weight gain was 2.39-3.29. It took 482 days from 150 to 450 kg, with a daily average input of ¥ 1.31, an output of ¥ 2.36 and net profit of ¥ 506.1 per animal.

*Table 5. Feeding experiments with ammoniated straw in Shangshui county*

| *Body weight (kg)* | *Concentrate (kg/day)* | *Straw intake (kg/day)* | *Weight gain (g/day)* | *Concentrate per kg gained* |
|---|---|---|---|---|
| 150-250 | 1.5 | 7.0 | 628 | 2.39 |
| 251-350 | 2.0 | 8.9 | 629 | 3.18 |
| 351-450 | 2.0 | 11.4 | 608 | 3.29 |

*Table 6. Economic analysis of studies in Shangshui country*

| *Body weight (kg)* | *Weight gain (g/day)* | *Inputs (¥/day)* | *Output (¥/day)* | *Benefit (¥/day)* |
|---|---|---|---|---|
| 150-250 | 628 | 1.05 | 2.39 | 1.34 |
| 251-350 | 629 | 1.38 | 2.39 | 1.01 |
| 351-450 | 608 | 1.50 | 2.31 | 0.81 |

In hebei province some researchers conducted a study on animals with straw sources. Twelve 18-month-old cattle were divided into 4 groups receiving ad libitum four straw sources: wheat straw, maize stover, ammoniated wheat straw and ammoniated maize stover. In addition, each animal received 1.5 kg of cotton seed cake per day. The study lasted 80 days, starting 21 February 1992. The results are presented in Table 7. The results confirmed that gains were better with ammoniated straw. Treatment of wheat straw improved growth 85.1 percent and treatment of maize stover by 45 percent. It also reduced the concentrate needed per kg of gain.

In other words, 1.99 kg concentrate was saved and straw conversion rate increased by 46.1 percent. Corresponding figures for maize stover were 2.93 to 2.02 kg of concentrate needed per kg gain; 0.91 kg of concentrate saved; and conversion improvement of 31.1 percent. Intake of crop residue DM varied from 2.14 to 2.83 kg per 100 kg of weight. Intake order was:

ammoniated maize stover > maize stover > ammoniated wheat straw > wheat straw.

*Table 7. Results of feeding experiments in Hebei Province*

| | | *Group* | | | |
|---|---|---|---|---|---|
| | | *Untreated wheat straw* | *Untreated maize stover* | *Ammoniated wheat straw* | *Ammoniated maize straw* |
| Number of cattle | | 3 | 3 | 3 | 3 |
| Days of experiment | | 80 | 80 | 80 | 80 |
| Average initial weight | (kg) | 234.8 | 234.8 | 234.8 | 234.8 |
| Total weight gained | (kg) | 83.4 | 122.8 | 154.5 | 178.6 |
| Average gain | (g/day) | 348 | 513 | 644 | 744 |
| Cottonseed intake | (kg/day) | 1.5 | 1.5 | 1.5 | 1.5 |
| Total feed intake | (% BW) | 2.56 | 3.10 | 2.94 | 3.29 |
| Straw intake | (% BW) | 2.14 | 2.70 | 2.48 | 2.83 |
| Concentrate: gain ratio | | 4.32 | 2.93 | 2.33 | 2.02 |
| Straw degradation rate | (%) | 45.03 | 62.97 | 54.75 | 69.31 |

Experiments were carried out on demonstration farms. Thirty local yellow cattle of 200 to 300 kg BW and 1.5 to 2 years old were selected and divided into 3 groups. Cattle were mainly fed *ad libitum* with ammoniated wheat straw. The groups were supplied with 2, 3 and 4 kg cottonseed cake, respectively, to study the effect of supplement level. The experiments started on 26 May 1992 and lasted 42 days. Results are presented in Table 8.

Results showed that daily gains increased with concentrate level, but in a diminishing response. From 2 to 3 kg, gains increased by 217 g/day, or 22.5 percent, but from 3 to 4 kg, only by 105 g/day, or 14.1 percent.

*Table 8. Effects of cottonseed cake ration on daily gain*

| | | Group | | |
|---|---|---|---|---|
| | | *I* | *II* | *III* |
| Number of cattle | | 10 | 10 | 10 |
| Experimental period | (days) | 42 | 42 | 42 |
| Cottonseed cake intake | (kg/day) | 2 | 3 | 4 |
| Ammoniated straw intake | (kg/day) | 7.25 | 7 | 4 |
| Total initial group weight | (kg) | 2 629 | 2532 | 2 847 |
| Total final group weight | (kg) | 28 85 | 2943 | 3 205 |
| Total weight gain | (kg) | 256 | 312 | 358 |
| Average daily gain | (g/day) | 609 | 826 | 931 |
| Concentrate to gain ratio | | 3.28:1 | 4.04:1 | 4.70:1 |
| Straw to gain ratio | | 11.89:1 | 9.44:1 | 7.93:1 |
| Cost of gain | (¥/kg) | 2.94 | 2.97 | 3.11 |
| Total income | (¥) | 922 | 1 121 | 1 287 |
| Total expenses | (¥) | 528 | 718 | 911 |
| Net benefit per head | (¥) | 39.4 | 40.4 | 37.6 |

The efficiency of concentrate use for growth also decreased with the level. The results from the above studies show that on the basis of ammoniated straw diets, the concentrate required per kg of weight gain varied from 0.83 to 3.5 kg, starting at 150 kg. The concentrate used per kg of liveweight gain was less in cattle than in swine. Cattle market age can be reduced to 2.5 years. This low-level concentrate feeding system is well suited for Chinese conditions.

## Ammoniated Straw for Dairy Cows

The Rowett Research Institute of UK reported the use of ammoniated straw for dairy cows instead of dry hay, without effects on milk yield or quality. Most researchers have tried ammoniated straw feeding for beef cattle and growing cattle, but there have been very few studies on high yielding dairy cows. In 1999, Lu Xilei from the Beijing Dairy Institute conducted systematic experiments feeding high production cows with ammoniated straw, with great success. Some results are presented here.

### Ammoniated Rice Straw

Ammoniated rice straw replaced 30 percent of Chinese wild-rye hay in the ration, without negative effects on yield and composition (Tables 9 & 10).

*Table 9. Groups of dairy cattle for ammoniated rice straw study*

| Group | n | Number of calvings | Milking days | Milk production (kg/day) | Body weight (kg) |
|---|---|---|---|---|---|
| Treated straw | 30 | 2.7 | 167 | 24.1 | 573 |
| Untreated straw | 30 | 2.8 | 164 | 23.6 | 576 |
| P | | > 0.05 | > 0.05 | > 0.05 | > 0.05 |

*Table 10. Average milk production and composition when feeding rice straw*

| Group | n | Days | Milk production kg | Fat (%) (kg/day) | Protein (%) | Lactose (%) |
|---|---|---|---|---|---|---|
| Treated straw | 30 | 112 | 24.12 | 3.24 | - | 4.75 |
| Untreated straw | 30 | 112 | 23.39 | 3.51 | 3.19 | 4.80 |
| P | | | > 0.05 | > 0.05 | > 0.05 | > 0.05 |

### Ammoniated Maize Straw

Feeding experiments were carried out supplying half of the ration as either ammoniated maize stover (group A) or untreated maize stover (group B). The other half was Chinese wild-rye hay. A third group (C) received only

Chinese wild-rye hay. A third group (C) received only Chinese wild-rye hay. No negative effects on milk production and composition were found when ammoniated maize stover or untreated maize stover substituted half of the Chinese wild-rye hay, but obvious economic profits were obtained. In other words, ammoniated straw could replace part of Chinese wild-rye hay for high yielding cows, but whether it could be replaced completely still needs further experimentation. A Shuangqiao farmer obtained good results with total replacement of Chinese wild-rye hay by ammoniated straw.

### Sheep Feeding

In Norway, Nedkvitne and Maurtvedt conducted an experiment with pregnant ewes, feeding them with straw and concentrates. One group was fed with 0.6-0.8 kg forage and another with 0.8-1.0 kg ammoniated straw. Mean values of the 3-year experiment are shown in Table 11. Results showed that there was no obvious difference in the number of lambs born between the two groups.

*Table 11. Results of feeding ammoniated straw to pregnant ewes*

| *Parameter* | | *Forage group* | *Ammonia straw group* |
|---|---|---|---|
| Feedstuffs | (kg/day) | | |
| Forage | | 0.6-0.8 | |
| Ammoniated straw | - | 0.8-1.0 | |
| Silage | 1.5-2.5 | 1.5-2.5 | |
| Concentrate | 0.2-0.3 | 0.2-0.3 | |
| Lambs per lambing | 54/31 | 45/34 | |
| Ewe weight on 12 December | (kg) | 74 | 75 |
| Weight gain during pregnancy | (kg) | 8.7 | 10.7 |

*Contd. . .*

| | | | |
|---|---|---|---|
| Net weight gain after lambing | (kg) | 2.7 | 1.0 |
| Lamb birth weight | (kg) | 5.1 | 5.1 |
| Lamb weight gain in spring | (g/day) | 344 | 344 |
| Lamb weight in autumn | (kg) | 49 | 49 |

Weight increases during pregnancy for the ammoniated straw group were more rapid than in the untreated group, but weight losses were also greater, so weight balance was better with forage. There were no differences in lamb weight in autumn. In recent years, some institutes in China have also used ammoniated straw in feeding experiments. For example, Zhen Erying from Hebei Agricultural University used urea-treated rice straw to feed sheep, with satisfactory results.

# 7

# Utilisation of Crop Residues and Agro-Industrial By-products

The value of crop residues and agro-industrial by-products as animal feed becomes more important to the Indian sub-continent than to other regions because of the 'dry spells' regularly experienced by countries in this region. A dry season of 3–4 months with no green fodder for grazing is a common feature in this area. Cattle and buffaloes often lose condition and even succumb to death for lack of feed.

There is, however, an abündant supply of crop residues, partiucularly cereal straw during this period because the dry season normally coincides with the harvesting time of cereal crops. There is very little information on the actual availability and uses of crop residues and other agro-industrial by-products in the Asian region. A rough estimate of the availability of crop residues can be made using extraction rates (Table 1).

Rice straw is the predominant fibrous residue, available in large quantities. It is also the major residue that is utilised to any reasonable extent in animal feeding. Statistics suggest that rice and wheat straws constitute over 50% of the total forage fed to cattle and buffaloes in India. The utilisation of rice straw as an animal feed is very variable in Sri Lanka; availability of green forage and other feed resources being the predominant factor governing this variation.

*Table 1. Availability of some by-products in the Indian sub-continent*

| Crop | By-product feed | Approximate extraction rate % | Availability of by-product (million tonnes) |
|---|---|---|---|
| *Cereal crops* | | | |
| Rice | straw | 100 | 98.5 |
| | bran | 10 | 9.8 |
| Wheat | straw | 100 | 50.8 |
| | bran | 10 | 5.1 |
| Barley plant | straw | 100 | 2.0 |
| Maize plant | stover | 200 | 16.2 |
| | cobs | 40–50 | 3.2–4.0 |
| Maize grain | bran | 8–10 | 0.6–0.8 |
| | germ-meal | 16–18 | 1.3–1.5 |
| Millet | straw | 200 | 18.4 |
| Sorghum | straw | 200 | 22.0 |
| *Other crops* | | | |
| Sugar cane | bagasse | 12–151 | 27.4–34.3 |
| | tops | 15–20 | 34.3–45.6 |
| | molasses | 3–4 | 6.8–9.1 |
| Roots and tubers | waste | 35–55 | 7.1–11.2 |
| Pulses | meal | 60–70 | 7.1–8.3 |
| Fruits | kernel and waste | 50–80 | 12.4–19.8 |
| Coconut | meal | 35–40 | 2.2–2.5 |

In certain parts of the country, particularly in the Northern peninsula, almost all the harvested straw is used for animal feeding. While in the central hills straw is hardly used at all as a feed material. Using the average nutrient contents of some principal by-products, Mahadevan estimated that in quantitative terms, crop residues can supply a substantial part of the maintenance requirements of all ruminants in the Asian region. However, because of their low feeding value, they are not used in production rations to any significant extent.

Crop residues and many agro-industrial by-products have certain inherent disadvantages in that nutritionally, they have low digestibility and are deficient in nitrogen and in many mineral elements; they are physically resistant to comminution and may contain high amounts of indigestible lignin and silica. Low digestibility associated with low nitrogen content of the feed limits intake and animals on these diets alone are often in negative energy and nitrogen balance. It is therefore essential that these deficiencies are corrected when crop residues and agricultural by-products are used as feeds.

As pointed out by Greenhalgh, despite the theoretical simplicity of supplementing straw with nitrogen and minerals, feeding systems based on straw supplemented with feed blocks or small amounts of concentrate have not proved quite satisfactory. The most recent innovation is the development of gelled molasses/nutrient blocks. These are now extensively studied by researchers in the Philippines through an FAO/UNDP project.

Apart from supplementation, many new techniques have been developed to increase the efficiency of utilisation of crop residues and by-products of agriculture. Of the methods available for physical processing, grinding has given reasonable results when roughages are included in complete diets but it is uneconomic and impractical in situations in Asia. Furthermore, it has been suggested that grinding does not improve the ME value of the roughage to any significant extent.

Biological methods are still in the early stages of development and it may be some time before methods are devised for large-scale application. The most applicable way of improving the ME value of a roughage is by chemical treatment. It has often been mentioned that processing is an 'optional extra' while the correction of nutrient deficiencies is 'essential' when considering poor quality roughages for animal feeding. However, often they become complementary as treatments which increase digestibility and intake may generate a need for additional nutrients to supply the requirements of the rumen microbes as well as the host animal.

### Chemical Treatment of Roughages

The chemical treatment of roughages preceded mechanical treatment, with the work of German pioneers such as Beckmann and Kellner. Since 1964, there has been a widespread renewal of interest in chemical treatment, particularly in the developing countries, which have the highest potential for crop residues as ruminant feed. Following are some general observations:

- In a majority of experiments, both in vitro and in vivo, sodium hydroxide has been used as the chemical reagent for treatment.
- There is no doubt as to the effect of sodium hydroxide on digestibility and intake of a roughage. In general, digestibility increases between 10–20% units can be expected with intake increases of 30–50%.
- Because of increase in digestibility and intake, the intake of ME is considerably increased leading to improved production. Various levels of body weight increases have been recorded depending upon the level of inclusion of the roughage in the diet. For example, Pirie and Greenhalgh reported 0.7–1.2 kg of gain in beef cattle fed diets containing

60% treated straw. Similarly, Jayasuriya et al showed that milk yields of Friesian cows can be increased by 26% by feeding a diet containing 65% treated rice straw.

- The achievements in (2) and (3) above are diminished by some drawbacks such as high price of sodium hydroxide, the corrosive nature of the chemical and the necessity for protective equipment and trained personnel for application. This has made sodium hydroxide treatment unfavourable amongst the Asian farmer.
- Ammonia is an alternative to sodium hydroxide for treatment of crop residues. It has been adopted in many European countries because (a) it does not leave residual alkali like sodium hydroxide and (b) it also increases the nitrogen content of the material. The standard method of ammonia treatment is not feasible under small farm situations in Asian countries because of the high cost of liquid or gaseous ammonia, lack of storage and transport facilities and lack of trained personnel to carry out treatment.
- One of the most promising methods of ammoniation is to use fertiliser grade urea as a solution and preserve the sprayed material under air tight conditions for 3-4 weeks. The method developed by Jackson in India and tested and modified by animal nutritionists in this region has clearly demonstrated that this treatment increases digestibility and intake of the residue leading to increased animal production. On-farms trials in Indai, Sri Lanka and Bangladesh have shown that urea-ammonia treated straw offers a great promise for the future of animal production in the Indian

sub-continent. These studies have also stressed the necessity for a simple and economic treatment system for the village farmer.

### Materials Needed for Straw Treatment

Ammoniation using a 4% solution of urea is a simple process that can be carried out at the village level. The appropriate quantity of urea dissolved in water is sprinkled on the straw either using a garden watering can or a tin can with many holes in it. Application need not be uniform as the ammonia formed on hydrolysis will penetrate into the moisture in the straw. The sprayed material is then stacked. An air tight cover is preferable to prevent the less of gaseous ammonia, but is not essential.

However, the feeding value of the treated material will vary according to the type of cover used and the choice of treatment method. As one can see, the method does not require special equipment or trained personnel to implement it. A careful examination of the method of treatment suggests the necessity of scales to weight straw and urea and means of measuring the required quantity of water and of spreading it as evently as possible.

Under practical conditions a farmer can be trained to bundle straw in approximately 10 kg lots. He can be taught to measure urea in approximate quantities using a pre-calibrated container. The same principle can be applied to measure water. Therefore, in practice, what is required from the point of view of equipment is only a bucket which is normally available in any household.

### Treatment Systems

The 'open stuck' method seems to be the most popular method of treatment in India while in Bangladesh straw is often treated in bamboo baskets covered with banana leaves or in pits dug in the ground. In Sri Lanka we have developed and tested three treatment systems, each having its own advantages and disadvantages.

***Closed System***

A pit 165 cm × 165 cm × 75 cm (length × width × depth respectively), partly buried in the ground, is built with bricks and cement. It is divided into four equal compartments by cross walls and the inside of each compartment is lined and smoothed with cement (Figure 1a). Straw treatment is carried out inside these compartments, each of which can hold approximately 60–70 kg of untreated straw, which is sufficient to feed 2–3 adult cattle (300 kg live weight) for 3 days. Straw is treated for 9 days according to a schedule (Table 2a) at 3 day intervals.

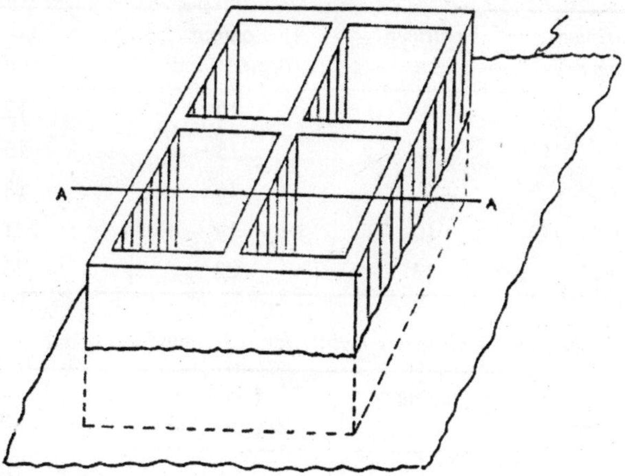


*Figure 1a. A four compartment pit made of brick and cement. (165 cm × 165 cm × 75 cm)*

As seen in the schedule, compartment 1 should be empty on the 13th day for the scheduled fifth treatment and this procedure should ensure a steady supply of treated straw. If the farmer considers the above programme cumbersome, he may have a two compartment pit of the proportionate size.

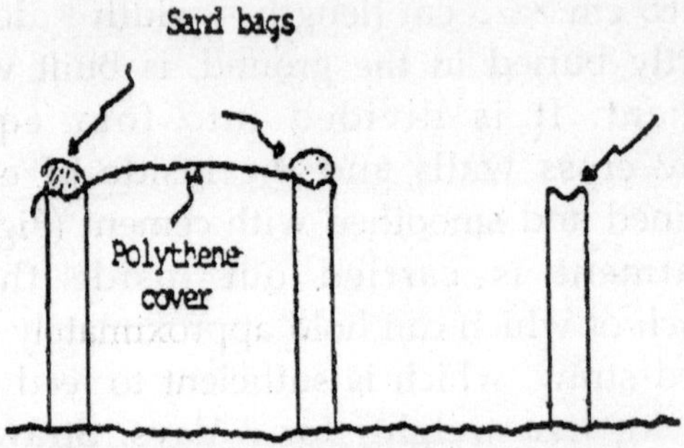


*Figure 1b. Cross section A-A showing the shallow groove on the pit walls*

*Table 2a. Treatment schedule for a 4 compartment pit*

| Compartment number | Treatment day | Feeding Begins (day) | Feeding ends |
|---|---|---|---|
| 1 | 1 | 10 | 12 |
| 2 | 4 | 13 | 15 |
| 3 | 7 | 16 | 18 |
| 4 | 10 | 19 | 21 |
| 1 | 13 | 22 | 24 |

*Table 2b. Treatment schedule for a 2 compartment pit*

| Compartment number | Treatment day | Feeding Begins (day) | Feeding ends |
|---|---|---|---|
| 1 | 1 | 4 | 10 |
| 2 | 8 | 11 | 17 |
| 1 | 15 | 18 | 24 |

Here he will treat every 8th day and feed the treated straw for an average of seven days starting from the 4th day after treatment (Table 2b). After treatment each compartment is covered with a polythene sheet and air tightness is ensured by having sand bags kept on the shallow grooves on the pit wall (Figures 1b).

Initially, straw was treated for 3–4 weeks, but laboratory investigations suggested very little benefit by extending the treatment period beyond 10 days. This was also found to be convenient from the point of view of the farmer.

#### Open System

The open system does not require a pit but a roof cover made of straw or cadjan is recommended. The required quantity of straw is sprinkled with the urea solution and the treated straw is stacked into a heap under the shed. A somewhat similar schedule to the one used for the closed system can be adopted but it is recommended that the treatment duration be less than 10 days as otherwise moulds tend to exceed the tolerable levels. Feeding of treated straw can begin by the 7th day after treatment so that feeding of that particular heap is complete in 10 days.

#### Semi-Open system

A shed having a roof cover is divided into four compartments of appropriate size according to requirements using either bamboo sticks or any other suitable material as shown in Figure 2. Straw is treated in batches inside each compartment according to a schedule similar to the one used for the open system (Table 2) and stacked into heaps. At the end of treatment both the sides and the top of the compartment can be covered either by jute hessian or used urea bags. Even green material such as banana leaves and coconut fronds spread over the heaps can help in minimising losses of moisture due to evaporation. The treatment period should not exceed 10 days as moulds tend to develop and make the material unpalatable.

Apart from these three treatment systems, straw can also be sprayed or sprinkled with a urea solution and fed immediately without having to preserve it. This is supplementation rather than treatment but it is possible under farm conditions.

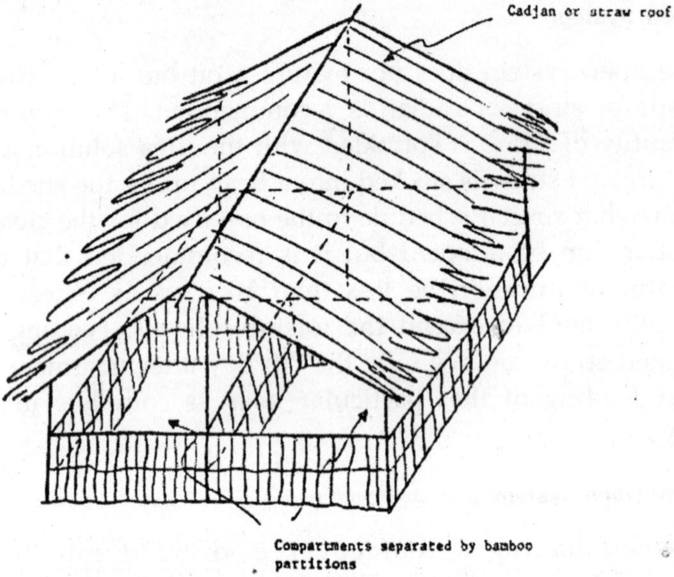


*Figure 2. A General view of the semi-open bamboo shed*

In all aspects the semi-open system is intermediary to the other treatment systems. The closed system is the most expensive as it requires initial capital investment. But it gives the best quality treated material with possibilities of higher animal production. The cement pit could also have other uses. It could be used as storage space for grain or water when straw treatment is not possible for various reasons. It could also serve as a silage pit for conserving pasture during the lush season.

*Table 3a. A comparison of some treatment systems using urea as a source of ammonia. Experiment with growing Sahiwal heifers.*

| | *Treatment* | | *System* |
|---|---|---|---|
| | *Supplemented straw* | *Semi-open system* | *Closed system* |
| Weight gain/day (kg) (by linear regression) | 0.169 | 0.240 | 0.319 |
| Straw DMI (kg/100 kg body weight) | 1.86 | 2.32 | 2.32 |
| Feed conversion ration (kg feed/kg gain) | 23.0 | 20.3 | 15.0 |

*Table 3b. A comparison of some treatment systems using urea as a source of ammonia. Experiments with lactating Surti buffaloes.*

| | *Treatment systems* | | |
|---|---|---|---|
| | *Supplemented straw* | *Open system* | *Closed system* |
| *Experiment 1:* | | | |
| Milk yield/ day(kg) | 2.40 | 2.95 | - |
| Butter fat production/day (g) | 173 | 231 | - |
| Straw DMI (kg/100 kg body weight) | 2.22 | 3.06 | - |
| *Experiment 2:* | | | |
| Milk yield/day (kg) | - | 2.67 | 2.94 |
| Butter fat production/day (g) | - | 212 | 231 |
| Straw DMI (kg/100 kg body weight) | - | 3.20 | 2.79 |
| Feed conversion ratio (feed/milk) | - | 1.44 | 1.17 |

The increase in digestibility and intake of a by-product due to treatment generates the need for extra nutrients for both rumen microbes and the host animal. This leads to the necessity of supplementing the treated material for increased animal performance.

## Supplementation of Crop Residues

Apart from alleviating nutritional deficiencies, an ideal supplement should also maintain or increase intake of the basal dietary material. The supplement should increase the efficiency of utilisation of nutrients leading to increased animal production. It is not always that a feed added to a basal diet acts as a true supplement. Instead of truly supplementing the nutrients in the basal diet, the added feed may substitute part of the nutrient supply.

The nutritional principles underlying the use of non-protein nitrogen (NPN) and rumen undegraded dietary protein (UDP) have been extensively discussed. It appears that in general while NPN supplies dietary requirements, UDP increases the amino acid supply to the animal, providing for increased production. Tree fodder legumes such as Glyricidia maculata and Leucaena leucocephala have been found to be highly suitable as supplements for treated straw diets. They contain a relatively high content of UDP compared to common protein supplements such as coconut oil meal, which presumably supplies the extra amino acid requirement of the animal.

Rice bran, one of the under-utilised by-products abundant in Asia, has been shown to be highly beneficial when added to straw based diets. Urea-ammonia treatment offers an excellent opportunity to increase animal production through crop residues and by-products of agricultural in the Indian sub-contient. It is a method highly suitable for the small farmer as it does not involve high technology and sophisticated machinery. Suitable supplementation with energy and protein should enhance its utilisation for production purposes.

The recent trend appears to be the supplementation by way of 'catlytic' quantities of both soluble and rumen non-degradable energy and protein. The soluble component is expected to stimulate the activity of rumen microbes enabling them to cope with the increased intake of the roughage. The rumen non-degradable component will then provide the extra energy and protein needed for growth and production. Any new feeding system must be economically viable to be acceptable to the farmer.

There is very little information on the economic feasibility of feeding urea treated straw to cattle and buffaloes. The cost of preparing treated straw can be easily calculated but response to treatment has not been established in actual monetary terms. One must realise that treated straw is not a substitute for good quality fodder. It is an emergency feed for periods of fodder shortage. Sustaining animal production during such periods and preventing death due to starvation has unlimited economic advantages. Straw treatment must be considered in this perspective.

### Availability and Current Utilisation of By-products

There is no doubt that crop residues, particularly cereal straws, are the major source of feed for cattle and buffalo in the Indian sub-continent. However, the picture is not sufficiently clear particularly in relation to the type and quantity of by-products fed and the seasonality of feeding. Ad hoc surveys have often pointed out the wide variation that exists in by-product utilisation between and within countries. Information is urgently required not only on the total availability of different by-products in each country but also on their availability in relation to the needs in different parts of the country.

Plenty of information is now available on the benefits of urea-ammonia treatment. Investigation should continue on the use of other chemicals such as calcium hydroxide for possible use in the future. The use of urea-molasses-mineral blocks with untreated residues in practical feeding systems should be throughly investigated, and if proved satisfactory,could be one of the simplest ways of increasing utilisation of crop residues in the region.

Special mention must be made of the necessity for further research in biological upgrading using fungi and extra cellular enzymes. The rapid development of genetic engineering will probably allow the manufacture of ligninases in the future. If economic, this could have major application for improving the nutritive value of crop residues.

Although some information is now forthcoming on the influence of supplementation of straw-based diets on voluntary feed intake, digestibility and production (milk, meat and draught power) more information is still required, particularly for by-products other than cereal straws. Sugar cane bagasse, cane tops, corn stover and corn cobs are some of the major by-products that need immediate evaluation. Supplementation should be by way of small quantities of green forage, tree fodder legumes (Glyricidia maculata, Leucaena leucocephala), industrial by-products (tea waste, oil seed meals), non-protein nitrogen (poultry litter) and minerals.

The effect of different levels of supplements on intake and digestibility of the basal roughage and on production parameters need monitoring. Nutritional studies should be accompanied by economic evaluation, particularly in relation to the small farmer and low levels of productivity.

Studies on the effect of plane of nutrition, using residue-based feeding systems, on reproductive performance and disease resistance need investigation. Improved nutrition will generally improve the reproductive performance of the animal. This may particularly be true with the water buffalo. Incorporation of straw and other crop residues in feeding systems suitable for the small farmer in the region needs further study. Investigations should be directed towards establishing maximum levels of inclusion of basal roughage for optimising production.

It is frequently claimed that the buffalo has a greater capacity to consume and digest low quality roughages compared with cattle. They are also credited with being more able to utilise non-protein nitrogen than cattle. However, the experimental evidence is not clear and there is need for more carefully conducted research. Similarities between species will allow extrapolation of nutritional information gained with cattle to buffalo and vice versa. Given the importance of the buffalo in this region, if they are shown to be better able to utilise fibrous residues than cattle, the evidence will promote the development of farming systems using by-products and the buffalo.

Judging from the research priorities for the region it is evident that rigorous research has to be done without delay. However, lack of qualified scientific personnel and facilities for research and extension are two of the major constraints that appear to limit research and development in this region. Poor remuneration and other hardships have driven many qualified personnel away from their homeland looking for greener pastures in the developed world. Training junior scientists through short-term, intensive training courses could be one way of overcoming the present shortage of personnel. A more permanent solution to the problem would be to persuade respective governments to provide better facilities and incentives to their scientific personnel involved in research and development.

# 8

## Safe Use of Plant and Animal By-Products

Around the world, there are enormous variations in approaches to the safe use of any feedstuff, including by-products. This is a reflection of the disease status, the economic status and the degree of public awareness of such issues in different countries. At the upper extreme are the countries of Western Europe and North America, which have very intensive animal industries.

In general, these countries have good animal disease status, populations that are relatively wealthy and a high degree of public awareness about animal product safety. As a result, the animal products produced in such countries are generally of high quality and represent a low risk to the consuming public. Such countries also have a high degree of regulation but waste considerable amounts of potentially high-quality material of relatively low risk to public health in order to try to eliminate all risk.

At the opposite extreme are the poor countries of Africa and Asia, which have largely extensive animal industries of poor health status, relatively poor populations and a low level of public awareness about animal product safety. From this description it is clear that the main issues relating to a country's human food safety include:

- the degree of intensification of its animal industries;
- wealth and its distribution within the country;
- public awareness of food issues;
- the degree of regulation of the animal production and animal feed industries;
- per capita consumption of animal products.

### Degree of Intensification of the Animal Industries

It may be expected that diseases are likely to be most common in intensive production systems. However, in most cases, diseases are generally better controlled in such situations than in extensive systems. This is principally a result of large intensive industries being able to justify high-tech facilities, more staff and better feeding and health management.

A major risk to public health from the intensive approach is the high level of drug usage that is adopted by such systems and that can result in:

- Contamination of animal by-products by the drugs used to control growth and diseases, including hormones, antibiotics, other non-antibiotic growth promotion products such as copper or arsenic, and preparations used to control internal and external parasites, which contain chemicals such as organophosphates.
- Antibiotic resistance caused by the overuse of antibiotics as growth promoters or to control the chronic persistent background diseases prevalent in such systems.

Another danger associated with intensive systems is the need to be continuously at the forefront of science in order to keep ahead of disease problems and attempt to increase profits. As a result, problems that previously were not experienced and expected will be encountered. The BSE crisis in the United Kingdom and, to a lesser extent, other European countries is one of the most significant examples of this. The disease appears to have developed as a result of changing the processing procedures of meat- and bonemeals in order to increase the speed of processing and reduce costs while, at the same time, using animal by-products in dairy cattle feeding in order to increase milk yields.

### Wealth and its Distribution within the Country

Wealthy countries generally have high per capita consumption of animal products and the poorest countries generally have the lowest consumption. For this reason, the issue of animal feed safety could be academic for many of the poorest people in the world, while the most wealthy are likely to consume only the highest-quality products, which are also likely to be the safest. It is therefore likely that animal feed safety will become more of a problem for middle-income people who are able to buy significant amounts of animal products but only those of lower quality and lower cost; and these are likely to be of the greatest risk.

### Awareness of Food Problems

Well-informed societies have become very focused on food issues, to the point at which concern for the dangers of food safety has risen to the limits of scientific possibility. As an example, the BSE crisis in the United Kingdom has resulted in all animal products now being banned from use as animal feeds. This includes materials of pig origin, which so far have not been shown to have any link to BSE, and materials that have been approved for human consumption. At the opposite extreme, less-informed societies could become the recipients of products and technologies that are not accepted in better-informed locations. Clearly, protection should be given to the latter, to prevent them from becoming either testing grounds for new, unproved technologies or recipients of products of doubtful health status.

### Regulation of the Animal Production and Animal Feed Industries

There is considerable variation in the regulation of the animal production and animal feed industries in different countries. While it is unlikely, and possibly inappropriate, that a common set of international standards be established, many countries are looking to develop and implement regulations that are applicable to their own circumstances. For example, the most developed countries have established very complete and in-depth sets of regularly upgraded regulations, based on many years of experience and designed to meet their own particular needs and circumstances. It would be very useful to produce a set of general guidelines on the development and establishment of regulations, which countries could adapt to their own circumstances.

### Per Capita Consumption of Animal Products

For reasons of wealth, culture and religion, many people around the world do not eat animal products. Clearly, for them this subject is irrelevant. However, since the non-consumption of animal products is often governed by lack of wealth, it is likely that such individuals could be at risk should their circumstances improve and the purchase of animal products become possible. Within this scenario the use of by-products in the feeding of animals is likely to be the main cause of risk.

***Plant By-products***

Plant by-product materials can be a potential danger to consuming animals owing to the presence of a wide range of materials including inherent toxic factors, contaminating compounds and pathogens and, possibly, genes from GMOs. Many plants contain toxic materials or compounds that have an antinutritive effect, such as protease inhibitors, phytohaemoglutinins, gossypol, tannins, phytic acid, saponins, alkaloids, cyanogens, lathyogens, cyclopropene fatty acids, erucic acid, phyto-oestrogens, allergens, toxic amino acids (mimosine, canavanine,) antivitamin factors and anti-enzymes.

In order that plant materials can be fed to animals it is important that these compounds are bred out of plants, or removed or deactivated before consumption. The following are some examples of toxic materials in plant by-products.

*Toxic compounds*: A good example of a toxic compound that is present in a well-used animal feed by-product is gossypol, which is a contaminant of cottonseed products. The use of cottonseed in animal feed has been facilitated by breeding plant types with lower levels of gossypol, removing the gossypol by physical extraction or deactivating it through the use of iron salts. In another case, that of erucic acid which is found in the seeds of such plants as rape or mustards, plants with low levels of the compound have been bred.

Many other plants contain high levels of toxic elements, such as heavy metals, resulting from their cultivation on soils with a high content of these elements. This subject was reviewed by Howell. Since animals and humans have varying tolerances to toxic plant materials, it is possible for the animals that consume them to show minor or no symptoms of their presence, but human health could be damaged. For example, pigs are much more tolerant to the presence of such elements as copper, which can accumulate in their livers to levels that damage the humans who consume the affected animal products.

*Antinutritive compounds*: Many plant materials contain antinutritive compounds of many types. Good examples of such products are the protease inhibitors or phytohaemoglutinins found in many leguminous plant materials. A cooking step is sufficient to deactivate such compounds and make the materials safe to use. Without such a simple procedure, the performance of consuming animals can be greatly impaired, although there is little risk that effects will be carried on to the humans who eat the animals.

*Contaminating compounds*: Plant materials can become contaminated with a wide range of chemical products, which can be dangerous to the animals that consume them. Examples of such chemicals include pesticides used to control insects or plant diseases, or chemicals from poorly cleaned stores and transporting vehicles. While many modern pesticides are not persistent and are of little danger when used as prescribed, there is the risk that specified usage may not always be followed and that regulations vary from country to country.

There is also danger in the use of those persistent pesticides that are still permitted for certain specific problems for which no non-toxic alternatives have been developed, for example, in the control of such cotton pests as bollworms. Unless the recommended use is strictly followed, these compounds may be consumed by animals eating by-products. Some persistent pesticides can be passed on and accumulate in animals further up the food chain, as is the case with organo phosphates, for example.

Plant by-products frequently contain the external coverings of the original plant products, such as skins, peals and brans. These are more likely to be contaminated than core material and, as a result, any contaminant is likely to become concentrated in by-products. Frequently, stores and transporting vehicles are used for a wide range of materials.

Plant products and related by-products can be contaminated by the residues of previously stored or transported materials, unless such practices are strictly regulated. In addition, the surfaces of stores and many plant materials are often treated with insecticides to control storage insects. For all of these reasons, there is probably a greater risk of plant by-products being dangerous to animal and public health than of the original unprocessed plant product.

*Contaminating pathogens*: The most significant risk to animal health is probably that resulting from the contamination of animal feedstuffs with pathogenic microbes or their toxins. The most important of these are the mycotoxins, which occur on many plant products and especially those of tropical origin. Mycotoxins are secondary metabolites of fungi whose toxic effects cause a diverse range of disorders in animals and humans.

In some cases they are acutely toxic and in others they are linked to certain types of cancers. The most commonly reported problems are associated with fungal contamination of grains, forages and oilseed cakes such as cottonseed and groundnut cakes. The *Aspergillus* spp. and *Fusarium* spp. of fungi, which produce several types of aflatoxins and fumonisins, respectively, are two of many well-documented examples of mycotoxins that occur predominantly on such materials in the tropics.

Reviews of the mycotoxins that are likely to occur on feeds and forages were recently produced by D'Mello and

Macdonald and Scudamore and Livesey, who also describe the main types and effects of mycotoxins on animal health. These authors report that recent research indicates promising prospects for the development of plant genotypes that are immune or resistant to infection by toxigenic fungi. They also state that, in the case of aflatoxin-contaminated oilseed meals, treatment with ammonia is a cost-effective method of detoxification.

Nevertheless in most situations the main approach to controlling the problem is to reduce contamination of plant products through careful field management and storage and to monitor levels of contamination on potentially contaminated materials before use.

*Genes from GMOs*: There is ever-increasing use of genetically modified plants and microorganisms around the world. This has been a result of the enormous potential of genetic manipulation as a way of making plants more disease- or pest-resistant. Technology also allows the nutritional composition of modified organisms to be changed and enables them to be used to produce a wide range of products such as enzymes, hormones and pharmaceuticals. As a result of this technology, it is very likely that large quantities of by-products originating from genetically modified materials will be ingested by the world's animal and human populations and there are possible dangers to animal and human health from their use.

This subject was recently reviewed by Brufau and Tacon, who emphasise the necessity of exploring the safety issues of genetic modification technology fully. In particular, they draw attention to the possible direct toxicity or antinutritional effects of transgene products, the indirect or unplanned effects on gut microbial ecology and metabolism and the possibility of onward gene transfer of transgenes to the resident microflora of the gut.

The science of genetic modification is in its infancy and current assessments of risk can only be based on the limited scientific information that is available. Until there is considerably more information about the use of these products, and the longer-term implications of their use are established, there will continue to be considerable concern about the safe use of the technology in both public and scientific circles. In the meantime, regulatory agencies in countries where genetically modified products are produced or used will need to monitor and regulate their use within proven known safety standards.

### Animal by-products

The following are the main by-products from animal sources that have been focuses of concern for animal and public health:

- rendered meat- and bonemeals;
- acid-preserved fish and carcass wastes processed by using added acids and acid produced by fermentation;
- swill containing restaurant and canteen wastes, waste foods and waste pig and poultry carcass material;
- faecal wastes and gut contents.

*Meat- and bonemeals*: The practice of feeding rendered animal proteins to other animals has been followed throughout the world for the past 50 years. The rendering process involves the cooking of parts of animals not used in human consumption at high temperatures to make the material safe and reduce the water content to a storable level. This process is then followed by mechanical separation of tallow (fat) from the solid residue, which is then ground to form meat- and bonemeals (MBMs). The materials used include fallen animals and all other parts of animals and poultry not used in human food which, because of the BSE problem, the United Kingdom is currently not permitting the use of in animal feeding.

Instead, these materials are rendered and put into storage for later incineration; at present, 400 000 tonnes are in storage across the country. In other parts of the world, however, MBMs continue to be used in animal feeds, and the standards set throughout Europe in response to the BSE crisis are of particular interest as guidelines for others. Research carried by the European Commission (EC) has found that the only rendering system that reduces the level of the scrapie agent—a prion held responsible for BSE and the human disease Creutzfeldt-Jakob disease (CJD)—to an undetectable amount is that of heating to a temperature of 133°C at 3 bar pressure for a minimum of 20 minutes. This system has been used to set new EC standards.

Several reports have concluded that the BSE problem first occurred in the United Kingdom because of a combination of critical factors peculiar to that country, including the existence of a sheep population in which scrapie occurs, a sheep population that is much larger than the cattle population, and a feed and cattle industry that produced and used ruminant-derived MBM in calf rations. Many other countries feed ruminant MBM to other ruminants, but BSE did not result because of the absence of the first two critical factors.

MBM is considered a very valuable animal feed product in most parts of the world. In the preparation of this article, 94 references describing the processing, composition and use of MBM around the world over the past ten years were consulted. They describe the feeding of different MBMs to many species of domestic animal, including cattle, pigs, poultry and fish. In all cases apart from that of BSE in the United Kingdom and the rest of Europe, appropriately processed MBM was shown not to have caused problems.

In all circumstances, MBM performed to its compositional potential and is therefore likely to continue to be used in most of the countries where it is currently available. On the other hand, owing to the limited supply and high costs of MBM materials in developing countries, as well as the poor food conversion efficiencies obtained when they are used in ruminant feed, it is unlikely that they will be widely used in those areas. The United Kingdom and European experience might be considered as a unique example that adds to the wider knowledge of MBM use. One lesson that might be taken from this experience is the danger of feeding animal proteins to ruminants whose digestive processes were not developed to utilise such materials.

*Acid-preserved fish and carcass waste*: The processing of fish and animal carcass waste with acids has been practised for many years. This subject was recently reviewed for FAO by Machin, who showed that such waste materials can be preserved by using organic acids, such as formic acid, and mixtures of organic and mineral acids, such as sulphuric or hydrochloric acid. The acids preserve the materials, following comminution, through a combination of autolysis that uses enzymes from the tissues being preserved and the effect of dissociation of organic acids into cells within the mixture, thereby killing pathogens.

However, this system has not been adopted as widely as might have been expected, largely as a result of the high cost of acids and the difficulty of handling strong acids on farms in developing countries. For these reasons, the process has mostly only been applied to the preservation of fish and carcass waste for use in fur or fish farming, which are less cost-sensitive. The alternative system for preserving such materials is to use natural fermentation to generate lactic acid.

The materials to be preserved rarely contain enough suitable carbohydrate to act as a fermentative substrate, so preservation by this technique requires that a suitable fermentative substrate be added to the animal or fish material to be preserved. Suitable fermentative substrates include easily fermented materials such as molasses or fruit wastes. Where materials rich in more complex carbohydrates are used it is often necessary to add a starter culture to initiate fermentation.

Such cultures have included lactic acid bacteria such as *Lactobacillus acidophilus*, or *L. plantarum* and *Streptococcus faecium*. In these cases, carbohydrate sources have included maize, wheat and wheat bran, and in one situation lactic acid was produced by fermenting straw with crab and fish waste to produce a silage for feeding to cattle. The resulting silages have been fed to pigs, poultry, fish (many types), sheep, goats and poultry.

The preservation of fish and carcass waste with organic acids is dependent on the ability of the acids to dissociate across semipermeable membranes, such as cell walls, and accumulate within the cells. In the case of bacteria and other pathogenic organisms, the drop in intracellular pH kills the organism; for example, *Salmonella* spp., *Clostridium* spp. and coliforms have been killed at pHs of less than $pH_4$. This research has had two interesting spin-offs: the direct addition of organic acids and their salts to animal feeds in order to act as growth promoters; and the fermentation of conventional feeds so that they produce lactic acid, thus acting as growth promoters. This work has been a result of the European ban on the use of antibiotics in animal feeds and the need to identify alternative means of controlling enteric pathogenic bacteria in animal production systems.

Overall, the use of acids in the preservation of animal and fish protein wastes has been successfully applied in animal production without problems. However, were it to be applied to the destruction of BSE agents, it seems unlikely that this approach would be any more effective than rendering is. Care should therefore be taken to ensure that ruminant protein materials processed with acids are not fed back to ruminants. In order to reduce the risk of diseases developing, acid-preserved fish and carcass waste should be fed only to monogastric animals that have already been fed this type of material and to dissimilar species.

As well as BSE-type agents, potential problems may occur when resistant parasite eggs from such nematodes as *Ascaris* spp. and bacterial spores are present in the materials and fed to susceptible animals. However, many of these species require an extra-animal maturation period before they can become infective and this could reduce the potential danger.

*Food processing waste (swill)*: The waste from catering establishments, including food manufacturers and in some cases raw meat producers, is used all over the world as an animal feedstuff. Such food processing waste is often known as swill and is used predominantly in pig and, to a lesser extent, poultry feeding.

In Europe and many other developed countries the use of these materials is very firmly regulated because of the danger to animal and human health that they can cause. Currently, the EC regulations regarding this area are being modified to take account of recent knowledge associated with BSE. In the United Kingdom, because of the BSE threat, no mammalian protein is permitted to be used as animal feed, although poultry and fish wastes can be used as swill as long as they are subject to processing that meets rendering standards. The regulations also prevent the feeding of poultry waste to the same species as a means of preventing disease transmission.

In Australia, swill feeding has been prohibited in order to prevent the introduction of exotic diseases. The regulations currently applied in the EC for swill processing involve cooking waste at 100°C for one hour, or at higher temperatures for proportionately shorter periods of time (down to 20 minutes in the case of 133°C at 3 bar pressure). The regulations also require that all the facilities involved in swill processing be subject to the surveillance of veterinary authorities. Such surveillance includes regular microbiological checks, separation of uncooked from cooked materials, and inspection and approval of all the premises that receive and use such materials. These regulations could usefully be used as guidelines for the establishment of regulations in other countries.

*Problems associated with swill feeding*: In fact, swill feeding has been shown to be responsible for the transmission of such diseases as swine fever and African swine fever. In many cases, the introduction of pig diseases into countries that were previously free of them has been associated with the use of inadequately processed swill produced from airport wastes or imported human foods.

Clearly, swill is a potentially very valuable animal feed raw material that would create environmental problems if not used as an animal feed. Nevertheless, it is also a potentially very dangerous product that, when not handled with care, has been shown to be a major cause of animal health breakdowns in several countries. It is therefore essential that all countries have effective processing regulations and enforcement procedures to control the handling of food processing wastes.

*Faecal waste and gut contents*: It is important to consider the use of poultry waste in this discussion since it is a valuable feed material in many countries, although it is not used as a feed in many developed countries because of public sensitivity about the feeding of faecal material to animals and the possible risks of disease transfer if the material is not fully processed. This subject was last reviewed by FAO in 1982. A considerable proportion of poultry faecal waste contains undigested nutrient material as well as non-protein nitrogen (NPN) that can be utilised by ruminants in fermentative digestion. In many countries, poultry waste has been extensively used as a valuable feedstuff for all types of livestock, including poultry, although predominantly to ruminant species.

In some cases the material has been fed without any processing, and in others it has been cooked, dried and ground or ensiled with a wide range of forage materials. While most references report no problems with feeding this material, some have reported disease outbreaks—including major outbreaks involving ruminants suffering from botulism—and problems when feeding faecal waste to preruminant calves arising from the hepatoxic properties of this type of material.

Faecal waste has also been fed extensively to many types of fish, including tilapias, carp and trout without problems. The fact that the fish did not appear to suffer hepatoxic problems may be because many fish actually maintain high levels of uric compounds in their blood or are able to discharge such materials with ease into the surrounding water.

*Processing*: From the literature, it would appear that faecal material should be heat-sterilised before being used as an animal feedstuff; although it seems likely that fermentation could be an effective process. Faecal material has been sterilised by high-temperature cooking followed by grinding. In one case, it was sterilised by heating at 133°C for ten hours. This temperature is in line with the rendering regulations applied in Europe, although ten hours would appear excessive. Material that has been ensiled also seems to have caused minimal problems. The effect of lactic acid has a sterilising effect. There is a need for up-to-date guidelines on safe processing and feeding of this material as both a dried heat-processed and an ensiled product. In particular, research is required, in order to confirm the effect of ensilation on the destruction of potentially pathogenic disease organisms.

*Use*: Because of the presence of non-protein hepatoxic materials such as uric acid it would appear logical to feed poultry faecal material to ruminant animals, which have rumen microorganisms capable of using this nutrient. Feeding to fish might have more potential than at first seems likely since any expelled NPN might act as a fertiliser to other aquatic plant species that can then become secondary feedstuffs. Although faecal material has been fed to poultry without apparent difficulty, it would appear to have only limited benefit and some risk of disease transmission even when heat-processed. From an environmental point of view it would seem highly beneficial to apply some form of processing to this type of material and then use it as an animal feed rather than use it for other purposes or dump it in an unprocessed form.

**Management of Plant and Animal By-products**

Over the past few years the importance of feed safety in the management of plant and animal by-products has become much more recognised. This has happened largely in response to the well publicised animal and human health problems that have occurred over recent years as a result of diseases associated with the use of these products. Although the most public disease outbreaks have occurred in Europe, the potential danger is a global issue and the limited health safety infrastructures in developing countries mean that problems there could go unrecognised. There is therefore a need for all countries to be aware of the problems with animal and plant by-product processing technology, and their solutions.

There is a need to identify the main issues and offer an easily accessible advisory package that will permit such counties to manage their own industries. Clearly, ensuring that materials are safe to use requires both long- and short-term approaches. In both cases, an in-depth understanding of the problems and ways to overcome them is essential. In the long term, such understanding will enable well thought out procedures for safe processing and treatment to be developed. In the short term, the feed users and regulating agencies need to understand the origins and methods of processing the materials so that monitoring systems can be established to confirm the levels and presence of any unsafe materials.

# 9

# Crop Residues as Feed Resources in Crop/Livestock Farming Systems

There have been a number of efforts to identify and describe farming systems in warm-climate regions based largely on geography (political and physical), climate, cropping pattern and animal output. Seldom has there been focus on crop-livestock relations. Emphasising crop-livestock relations, McDowell and Hildebrand identified prevailing systems on small, mixed farms in Africa, Asia and Latin America. Ten major systems were identified in Asia: with the exception of swidden (slash-and-burn) farming, crop residues and byproducts from human food processing provided 30 to 90% of livestock feed.

Africa has 10 major systems with 22 subsystems: dependence of livestock on crop residues was high in all 22 subsystems. In Latin America, four major systems were identified: in all except one, (commercial cattle ranching) crop residues provided 30 to 90% of livestock feed. Nearly all systems on the three continents also depended on grazing from fallowed crop lands.

Average farm size is approximately 1 ha and more than 85% of farms have livestock, usually two or more species. The interdependence of crops and livestock is primarily through dependence of animals on crop residues for feed. Farmers manage individual cropping patterns (intercropped maize and beans, double cropped maize and cassava) to provide food and feed. Each crop contributes feed during various months. Neither crop nor livestock productivity can be increased without due consideration of the interaction between crops and livestock.

Farmers keep, on average, one cow, two sheep and several poultry using farm and external feed sources, such as grazing or material cut from roadsides and forest. Maize stover is the most important feed. Farmers in this area have made little use of improved maize varieties because their stover yield is low unless fertiliser is applied and the indigestible neutral detergent fibre fraction (INDF) in their stover is higher whether with or without fertiliser.

Although crop and animal production can be strongly interdependent, the factors that can influence farmers' decision-making are often more complex (Figure 1). Sands made an in-depth study of the contributions of animals on 80 farms in two districts of western Kenya (mean size 1.03 ha). Using a two-dimensional model (household-market and household-farm), as proposed by McDowell and Hildebrand, two major subsystems requiring labour and capital were characterised. The solid lines from the crop subsystem to the animal subsystem show high dependence of animals on crop residues and strong dependence of cropping on animals for power in land preparation and fertiliser from manure dropped on fallow land while grazing or manure collected from night holding areas for composting with inedible crop residues.

The solid line between crops and the market indicates the importance of crop sales: grain sales provide over 20% of household income. The broken line from market

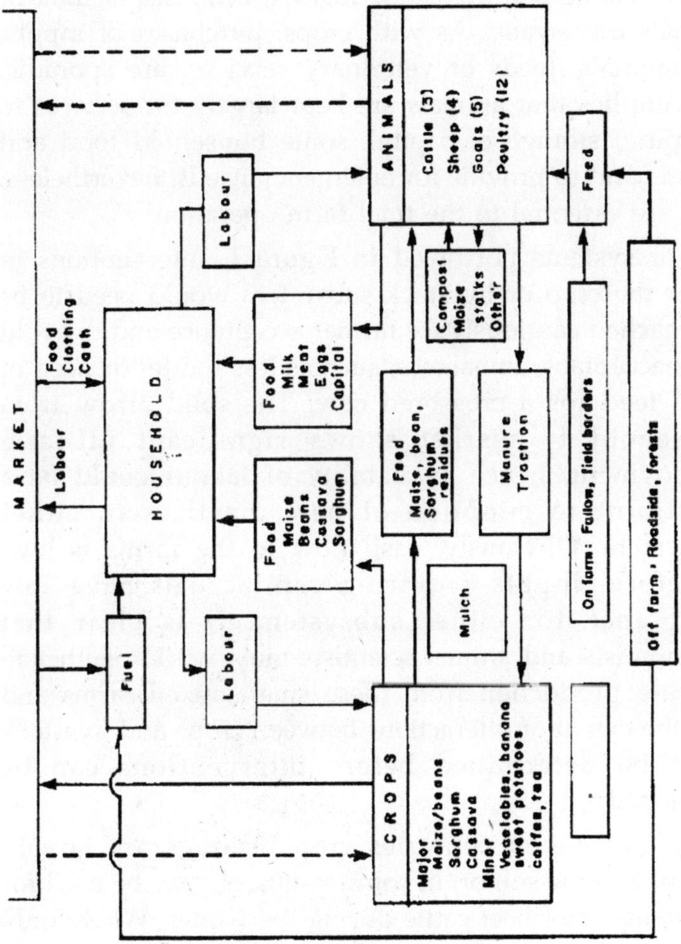


*Figure 1. Model of typical low resource farms in the highlands with crop and animal sub-systems.*

to crops shows low dependence on the market for inputs of seed, fertiliser or pesticides. Although annual income from sale of animals or their products equals or exceeds income from crops, the animal subsystem has an unpredictable relation to the market, i.e., sales of milk or animals are erratic. As with crops, purchases of inputs, e.g. animals, feeds or veterinary services, are sporadic. This implies that animals are kept largely for services to cropping, storage of capital, some household food and income and to provide for emergency needs: nevertheless, they are essential to the total farm operation.

For systems portrayed in Figure 1, interventions in either the crop or livestock subsystem would need to be approached cautiously for farmer acceptance and to avoid an unacceptable imbalance, such as less fodder or need of more feed for a crossbred cow. The solid arrow from household to market shows significant off-farm employment, hence availability of labour could be a constraint to adoption of new practices in either subsystem. Obviously, cash flow to the farms is low; therefore, inputs requiring capital will have low acceptance for either subsystem. It is clear that agronomists and animal scientists must work together to increase production from these small mixed farms and that the extent of interactions between crops and livestock must be determined before interventions can be developed.

In developed countries crop residues are largely returned to the soil or, in some instances, may be used for maintenance of beef cattle during the winter. Where only the grain is used, the overall efficiency of utilisation of total energy from a crop such as maize is low. One hectare of maize may yield approximately 30 240 Mcal of metabolisable energy (ME) and 620 kg of protein in the grain and stalk (Table 1). When only the grain is used for human consumption or for livestock feed about 39% of the energy and 20% of the protein are utilised. When the bran and stover are also used as animal feed, total ME utilisation may be increased to over 56% and protein utilisation to 28% (Table 1).

*Table 1. Production and utilisation of maize.*

| Production ha$^{-1}$ | Mcal ME | Protein (kg) |
|---|---|---|
| Grain | 19 040 | 360 |
| Plant | 11 200 | 260 |
| Total | 30 240 | 620 |
| Human or animal feed U.S. | | |
| Grain | 11 735 | 123 |
| % total | 38.8 | 20.0 |
| Subsistence farms Grain | | |
| Human consumption | 11 735 | 123 |
| Bran, animal feed | 2 810 | 22 |
| Stover, animal feed | | |
| Milk | 2 580 | 31 |
| % total | 56.7 | 28.4 |

Table 2 shows the high seasonal dependence of small farmers (1.3 ha) in southern India on crop residues (dry fodder). In this region milk sales are important, hence some purchased concentrates plus brans are used as supplements to maximise intake of the coarse feeds. Dry fodder provides 13% of feed dry matter from August to October, but 52% from January to April.

African pastoralists are highly dependent on crop residues from their own small plantings or from crop farms to supplement grazing during the dry season. Estimates of intake from grazing by 250-300 kg cattle are expressed in relation to animal needs for body maintenance (1.0). In northern Nigeria, grazing normally provides sufficient energy from June to October for a cow to gain up to 1.0 kg per day (July to September) but grazing cannot meet the animal's energy needs from December through May, resulting in serious weight losses.

*Table 2. Feed sources on mixed farms used for feeding buffaloes and cattle by season in southern India (kg animal$^{-1}$ day$^{-1}$).*

| | Jan-Apr | May-July | Aug-Oct | Nov-Dec |
|---|---|---|---|---|
| Green fodder | 2.22 | 2.20 | 9.06 | 6.19 |
| Dry fodder | 5.87 | 4.02 | 1.15 | 4.55 |
| Purchased concentrates | 0.89 | 0.19 | 0.40 | 0.44 |
| House concentrates | 1.51 | 0.65 | 0.40 | 0.16 |
| Other | 0.08 | 0.34 | 2.31 | 3.27 |
| Pasture (h d$^{-1}$) | 3.31 | 3.40 | 3.23 | 3.46 |

Fluctuations in feed quality and quantity lead to low net weight gains of about 70 kg a year. In Botswana the feed is deficient for only about 4 months but grazing and browsing in Botswana requires greater energy expenditure than in Nigeria. Even so, animal gain could reach 90 kg per year. If sufficient grazing is available, there is less need for supplementary feeding in Kenya than in Botswana or Nigeria. Expected animal gains would exceed 110 kg per year. Average rainfall in the three areas is approximately the same. Its distribution has a marked effect on the grass species and their nutritive value, which is highest in Kenya, somewhat lower in Botswana and least in Nigeria.

In Kenya and Botswana browse adds significantly to feed quality and quantity. In Nigeria, heavy rains over a short period lead to rapid growth and maturity of grasses followed by marked decline in quality. As pointed out by Wilson and others, supplementary feeding is essential in much of the subhumid and semi-arid areas of West

Africa. The need for manure on cropped areas and the pastoralists' need for feed leads to strong interdependence between crop farms and pastoralists.

An average pastoral unit requires about 10 breeding cows, a breeding male and associated stock for subsistence needs. These animals may use about 100 ha of rangeland providing approximately 10 500 kg of total digestible nutrients (TDN) per annum, which is far below needs. In northern Mali average rangelands in normal years will provide about 50% of livestock needs, hence crop residues must be used to avoid large weight losses in the dry season.

Data have not as yet been collected to determine whether the interrelationships between crop farmers and pastoralists in West Africa have influenced the adoption of new varieties of grain crops due to possible changes in yield and quality of crop residues, but ILCA researchers reported low acceptance of high-yielding varieties of cow-peas in northern Mali because of low forage yield. Concerns have been expressed in northern Nigeria and other areas over the rapid expansion of maize production, because maize matures earlier than sorghum or millet, while grazing is still reasonably good. As a result, the quantity and feeding value of maize stover is markedly lowered by weathering before it is needed for feed.

## Farmer Decision-making

### *Crop Selection*

Subsistence farms attempt to sustain about 4.5 people per household, each needing about 200 kg of grain per year. The farm must thus produce a total of 900 to 1000 kg of grain a year. Farm size is 1.5 ha, with 1.0 ha planted to maize and beans, 0.3 ha to wheat and 0.2 ha to sorghum.

Using local varieties and low inputs, maize yields 600 kg of grain, beans 150 kg, wheat 200 kg and sorghum 150 kg, giving a total yield of 1100 kg, about basic human food needs.

There are two cows, one bullock, one calf, two sheep and three goats. Yield of wheat straw is about 200 kg (1:1 ratio with grain) and maize plus sorghum stover 3450 kg (1:5 or 6 ratio to grain yield, in local varieties). Thus, the total crop residue yield is about 3650 kg, which provides approximately 150 days of feed. This, supplemented with off-farm grazing, could maintain the livestock. Cash flow is low so the farmers want to reduce the maize land to 0.5 ha and add 0.5 ha of cotton as a cash crop.

A new variety of maize is used and fertiliser applied, giving a yield of 1000 kg of grain. Of this, 200 to 300 kg is sold to pay for purchased inputs. The ratio of maize stover to grain yield is reduced from 1:5 or 6 to 1:1.5 or 2, hence maize stover yield is reduced to 2000 kg. The cotton provides no feed except weeds, hence total crop residue for dry season feeding is reduced to around 2500 kg, resulting in only 100 days of feed. The cattle fare less well because the digestibility of the maize stover declined from 52-56% (sufficient energy available for maintenance needs plus some for production) to 42-45% digestibility (sub-maintenance needs in energy).

The farmer must choose between reducing stock numbers, which is unattractive due to loss of prestige and savings, purchasing feed for livestock, relying more on off-farm grazing or returning to the traditional system. Other farmers have followed a similar procedure thereby placing greater pressure on communal grazing. In the second year nearly 50% of the farmers withdraw from the maize-cotton programme, to the consternation of extension agents.

***Management of Crop Residue***

Crop residues are low in protein and phosphorus, marginal in calcium and high in fibre and lignin. As a result, digestion is slow, rate of passage is low and voluntary intake is limited, e.g. ad libitum intake of sorghum stover is 43% less than that of hay. Intake may be increased about 20% by chopping the residue. Maize or sorghum may be cut and stacked or shocked to reduce leaf loss from leaching or wind damage.

Research has shown that stripping the lower leaves (below the ear on maize or the lower half of sorghum) increases feeding value. Topping maize after the grain has nearly matured also helps to preserve forage quality. Although these procedures improve feed quality, farmer acceptance has been low because of low visibility of return to the extra labour required. Assembling or storing crop residues may be a necessity where cropland is highly fragmented, such as often occurs in India; where the household is dependent on manure for fuel-India and Ethiopian highlands; or where marauding animals have access to crop residues during the off-crop season.

When these elements are not pressing, farmers prefer to graze the residues to reduce labour for storage or transport of manure to the fields. Preservation of crop residues is, however, attractive where high-protein concentrates, such as cottonseed cake or grain brans, are available at modest prices. With supplement, intake of residues may increase 20 to 30%. Farmers generally accept supplementation as an initial move to increase milk output or to fatten cattle or sheep.

***Chemical Treatment***

Chemical treatment could be more attractive if it were complemented with modifications in the farming system.

Forage crops have not received much attention in cropping systems research. Including forage legumes in the crop rotation can increase the yield of the subsequent crop and sustain soil fertility, and such rotations need investigation. Indirectly, forage legumes would increase returns from crop residues through higher intake and more efficient digestion. Availability of good quality forages for supplement may, however, lessen the attractiveness of chemical treatment.

#### Change the Animal

In recent years a frequent recommendation is for smallholders to concentrate on goats and sheep instead of cattle or buffalo. The advantages of low investment, early maturity and better breeding efficiency are most often cited. This recommendation makes the assumption that all four species are equally adept in the utilisation of crop residues, but this is not the case. Feeding strategy is an important feature in assessing suitability of animal species.

The comparative digestive strategies of goats, sheep and cattle are given in Table 3. Figure 2 portrays how selective feeding behaviour influences whether a given animal species is widely dispersed or clustered in certain areas because of prevailing feed resources. The two major types of Bubalus bubalis (swamp and riverine buffalo) are probably the best users of crop residues among domestic livestock.

Buffalo are grazers with low selectivity (Figure 2); they have a wide muzzle, large gut capacity and a greater extent of fermentation in the rumen than cattle. The last two features result in slow passage of food through the digestive system. The buffalo is therefore an effective user of high fibre feeds. Their major habitat, the paddy rice area of Asia, verifies their ability to use rice straw. Their best niche appears to be as users of crop residues or to provide some returns from grazing marsh areas. Their efficiency on high-quality forage is lower than that of cattle.

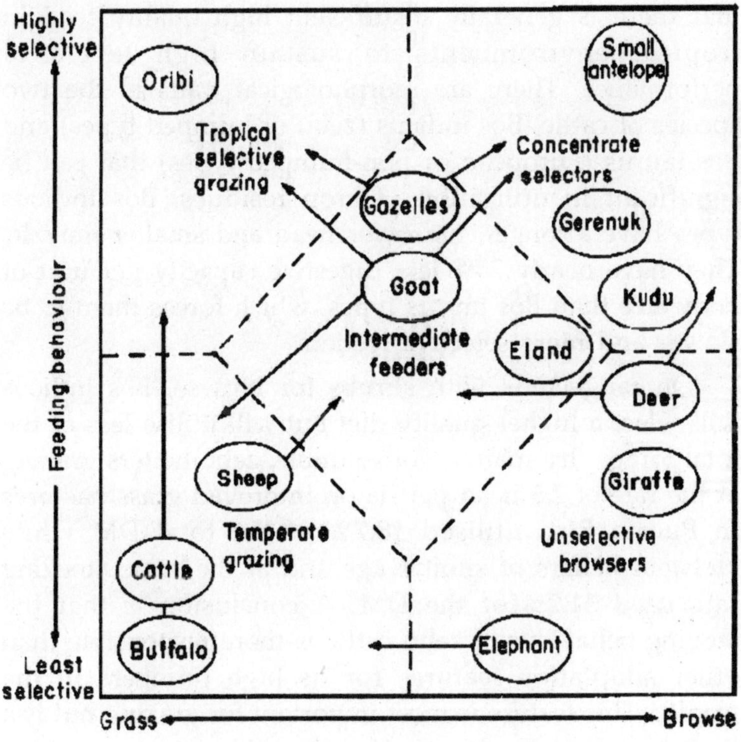


*Figure 2. Familiar and exotic animals are found based on their feeding behaviour and preferred feed resources.*

Cattle are classed as grazers with relatively low selectivity but are slightly more selective than buffalo. Their metabolic rate is lower than that of goats or sheep and their rumen retention time is longer, resulting in greater ability to digest fibre. The vast bacterial population in the rumen of cattle is a significant source of protein. Thus cattle can survive on a diet of poorer quality grazing than can goats or sheep.

Cattle can browse to a limited extent but their broad muzzle and slow bite rate does not make them effective browsers. Their absolute intake requirements are so great that there is generally insufficient high quality feed in tropical environments to sustain high levels of performance. There are morphological traits in the two species of cattle, Bos indicus (zebu or humped types) and Bos taurus (European or non-humped types) that can be significant in utilisation of crop residues. Bos indicus types have a longer, narrower head and smaller muzzle. They have nearly 25% less digestive capacity per unit of body size than Bos taurus types, which forces them to be slower and more selective feeders.

On rangelands with shrubs for browse, Bos indicus will select a higher quality diet but will utilise less of the total forage dry matter. For example, zebu heifers grazing at the rate of 2.5 head per ha on improved grass pastures in Puerto Rico utilised 18.7% of the total DM while Holstein heifers of similar age and at the same stocking rate used 31.2% of the DM. A conclusion is that the feeding behaviour of zebu cattle is more responsible than other adaptation features for its high numbers in the tropics. This feature is most important for grazing but is a limitation for zebu in use of crop residues.

In India, Pakistan and other parts of southeast Asia, a buffalo cow fed ad libitum rice straw will maintain body weight and produce 1.0-1.8 litres of milk per day while local cattle will lose weight. Goats are among the most selective of the intermediate feeders (Figure 2) and can use a wide range of plants. Their Mi (endogenous and microbial fraction of the faeces) values are lower than those of cattle and sheep (Table 3) because they generate less cellulytic bacteria in the rumen, which lessens cellulose digestion.

*Table 3. Comparative digestive strategies of goats, sheep and cattle.*

| Characteristics | Advantage | Limitation |
|---|---|---|
| | *Goat* | |
| Browser | Plant differentiation (morphological and seasonal) | Low utilisation of total biomass |
| Rapid passage | Higher intake possible, rapid passage of low-quality feed | More time required for eating |
| Low Mia | Allows greater allot selection requirements | High energy cost, increased maintenance |
| | *Sheep* | |
| Grazer | Less travel energy needs, better use of total biomass fully utilised | Diet limited to graze plants, plant differentiation not |
| Intermediate rate | Better fibre digestion of passage | Forced to waste effort on low-quality feeds |
| Intermediate Mi | Permits higher cellulose fermentation | Decreased apparent digestibility |
| Small body size | Small absolute intake | High MR/GC |
| | *Cattle* | |
| Grazer | High use of plant biomass | Diet limited to graze plants, plant differentiation limited |
| Slow passage | High fibre digestion | Forced to waste effort on low-quality feed |
| High Mi | High cellulose fermentation | Decreases apparent digestibility |
| Large body size | Low MR/GC, long legs, move rapidly | Mouth too large for plant differentiation |

The feeding strategy of goats is to select grasses when protein content and digestibility are high but to shift to browsing when leaves, bark and fruits have better nutritive value. Their small mouth and prehensile lips enable them to gather small leaves and flowers.

Performance of goats may be low and mortality high on an exclusive diet of dry-season grasses in the subhumid zone but they will thrive in the same zone where there is a mix of browse, while cattle may be hard pressed to survive. Overall, goats are not good users of straws and stovers unless they are given an opportunity for high selection and receive some supplementary protein as bran or browse. Goats may die on an all maize stover diet or when penned on dry, mature tropical grass. Thus, goats have unique feeding strategies which can be employed to complement sheep or cattle for fullest use of certain ecosystems, but where crop residues are the main feed source goats are at a disadvantage.

Sheep tend to be mainly grazers but, as for goats, their body size requires they feed selectively. They are able to digest fibre effectively but on a diet mainly of crop residues, such as straw, they have the disadvantage of being forced to ruminate in order to clear their rumen; therefore, straw or stover gives low nutritional benefit for the energy expended. It is more difficult to relate feeding behaviour to area of concentration for sheep than buffalo, cattle or goats because they have high utility for meat and fibre as well as importance as a feature of the Moslem religion. They usually complement other species in maximising use of ecosystems.

McDowell and Hildebrand showed that even though the number of small ruminants per farm was low, most had cattle, goats, sheep and, in Asia, buffaloes. This indicates that:

a) farmers are aware of the limitations of each animal species;

b) they know the complementarity of species in utilisation of available resources; and

c) they recognise that each species has a well defined function in the farm enterprise.

It appears that for much of Africa cattle will tend to dominate as the best overall user of feed resources but in planning agronomic or livestock research strategies, small ruminants should be considered as part of almost all farm systems.

# 10

# Crop Residues and Agro-Industrial By-Products in Developing Countries

To establish procedures for research into the treatment of crop residues and agro-industrial by-products in developing countries seems to be an extremely difficult task. The main reason being that conditions in the developing world vary so much from South East Asia to Africa to Latin America. Even within these regions extremes are found as to farming conditions, level of technology, infrastructure, economy and political system. These factors may have a fundamental influence on the strategy that should be chosen in an attempt to improve the utilisation of crop residues and agro-industrial by-products.

The livestock productivity in the tropics has not, in general, increased during the last 30 years, mainly because inappropriate technology has been promoted. Even if the explanation is not as simple as that, it is an extremely important statement to bear in mind for those who are involved in work on low quality roughage utilisation in developing countries. Therefore no research programme on treatment of low quality roughages should be launched before a feasibility study has been carried out. The study need not be very comprehensive, but it should make the planners think thoroughly through some questions that ought to be clarified before starting on a research programme. Before this is done, it is difficult to decide where to start and where to go.

### Quantities of Crop Residues

This is a logical point to start at because the quantities of crop residues and agro-industrial by-products will determine the attention that should be given to them as a potential source of feed. The possibility for building up a country-wide system for treatment will also depend on the actual amount available and its distribution. If grain production is located far from animal production, the possibility of utilising the straw as animal feed is greatly reduced. The transport cost of the straw may simply be prohibitive. Screening tests may be carried out to get valuable information about the degradability of the various materials. This is important in order to judge what sort of treatment or supplementation would be most suitable to improve its feeding value.

Several of the crop residues and by-products available today are used commercially for purposes other than feeding. Examples are sugarcane bagasse used as fuel, straw used as mulch, bedding, fuel, building material or in the chemical industry. One of the most common uses of straw is of course, to plough it under to improve the organic matter content of the soil. It is likely that technical developments will appear in the future which will make straw, bagasse, maize cobs etc. more attractive raw materials for industrial production. Such uses should be encouraged in areas where the crop residues and by-products cannot be utilised for animal feeding without long and expensive transportation.

It seems obvious that the incentive to treat crop residues and agro-industrial by-products is less in areas where high quality forages and concentrate feeds of low cost are abundant than in countries where feed alternatives to straw are scarce. It may be argued that if by-products could replace part or most of the grass products in the diet, more land would be available for production of human food. In many countries where this would apply, the market for human products is saturated and the world market prices do not pay for the extra work.

One interesting aspect of this question is the energy cost for production of various forages. Studies in Norway have shown that the expenditure of commercial energyper unit of metabolisable energy (ME) is much lower for ammonia treated straw than for grass silage, hay and other forages (Table 1).

*Table 1. The use of commercial energy in feed production*

| | MJ per MJ ME |
|---|---|
| Barley | 0.45 |
| Ammonia treated straw | 0.37 |
| Hay, wire fences + pasture | 0.51 |
| Hay, field cured + pasture | 0.54 |
| Pasture | 0.64 |
| Grass silage | 0.67 |
| Artificially dried grass | 2.10 |

The role of treated by-products in high concentrate diets is a matter of dispute. Because of the fermentation taking place in the rumen of animals fed such diets (low pH) the cellulolytic activity is greatly reduced. This is particularly severe if the materials are finely chopped or ground. Then the rate of passage is high and the time allowed for microbial degradation of the roughages becomes limiting. Kristensen, in reviewing the use of straw and other fibrous by-products in practical diets for cattle concludes that "ground straw is unsuitable in the diet of dairy cows". This point may not be so relevant in developing countries where straw and other by-products are fed in long form or coarsely chopped, and where the amounts of concentrates used normally are negligible. But, the point is worth noting particularly in situations where molasses and other readily fermentable carbohydrates are fed.

In most instances crop residues and agro-industrial by-products need to be supplemented with protein, minerals, etc. to ensure an optimal utilisation of the diet. These supplements should be available at a price that is not prohibitive for the small farmer. In many cases, urea is an interesing alternative to protein supplementation. On the other hand, catalytic supplements of undegradable protein to urea-treated straw may have a remarkably good effect.

**Influence of Technologies in Agricultural Sector**

Before initiating research into the treatment of crop residues and agro-industrial by-products, an evaluation of what is appropriate technology should be undertaken. What is appropriate may vary from one country to another. According to El Naga mobile straw treatment machines would be appropriate in Egypt. Parra and Escobar have described how sugar cane pith is treated with NaOH in Cuba. For other countries treatment with NaOH would be totally out of the question. Research on oven treatment of straw with ammonia in a country where most of the small farmers are illiterate, would be an example of a completely wrong approach.

The importance of advocating a technology that fits into the farming system and helps the farmer to get more out of his crop residues by simple means can never be over emphasised. The research in this area should primarily be devoted to problems the small-holders are facing. Research for the benefit of the feed mills and the feed industry should be given second priority. Firstly, because information obtained elsewhere may be adopted more directly here. Secondly, because the economy is generally better for the feed industry than for the small-holder in the countryside.

Another point is that when treated on the farm the transport costs are minimised. On the other hand, training of each individual farmer in a treatment method is a much greater task than training the technical personnel from the feed mills. The risk of accidents and false treatment is also greater when the farmers have to do the treatment themselves. Accidents that may occur are: explosion of anhydrous ammonia, burning of eyes and skin by caustic soda and poisoning of animals with urea.

## Effects of Treatment

It is possible to double the nutritive value of straw by alkali treatment. Ideally we may say that we choose the method that gives us the greatest improvement. There are, however, many other factors that ought to be considered before choosing one or two methods to be tested. Worth mentioning is that a ration of urea-treated rice straw plus 0.2 kg protein supplement and 0.2 kg green fodder has given a growth rate of 200 – 300 g/d for a 50 kg calf in Bangladesh. A similar diet gave 5 litres of milk per day. These performances are 2–3 times higher than the present averages in the country.

Studies in Norway showed that 400 kg steers gained 400 g/d for over 5 months on a diet of ammonia-treated straw plus 250 g of a concentrate mixture, minerals and vitamins. Studies by Mwakatundu and Owen showed that the improvement in digestibility by NaOH-treatment was greater for materials with an initially low digestibility than for materials with a high digestibility. This does not necessarily imply that highly lignified materials become highly digestible after chemical treatment.

Experiments with rape straw, bagasse, cotton stalk and rice brans indicate that such materials will have low digestibility even after chemical treatment. It may also be that weak alkalis such as $Ca(OH)_2$ or ammonia are unable to improve the digestibility whereas wet treatment with NaOH gives a satisfactory improvement in digestibility. Table 2 shows the in vitro, in sacco and in vivo digestibility of barley straw treated according to various methods.

It also shows the time taken for half the material to be degraded in the rumen. In the digestibility experiments with sheep 75 g of herring meal was added as a protein supplement and the digestibility of the straw was calculated by difference. The results clearly show that when sufficient protein is provided, there is no improvement in the digestibility by adding urea to the diet.

Urea and urine treatment for 8 weeks improved the digestibility of the straw, particularly when a small amount of soybean meal as a source of urease was added. Treatment of straw with anhydrous ammonia (stack) or sodium hydroxide increased the in vivo digestibility by 14–16 per cent units. The effect of aqueous ammonia and oven treatment with anhydrous ammonia was slightly less in these experiments. Highest digestibility was obtained for wet treatment of straw which improved the DM digestibility by 20– 25 per cent units.

*Table 2. Digestibility (%) of untreated and treated barley straw determined by different methods*

| Treatment | IVDMD (48 h) | In sacco dry matter degradability | In vivo DM digestibility | ±SEM | $T_{1/2}(h)$ |
|---|---|---|---|---|---|
| Untreated straw | 42.2 | 41.4 | 50.8 | 3.0 | 45.6 |
| Untreated straw + urea (at feeding) | | | 49.3 | | |
| Urine treated straw | 58.5 | 56.2 | 56.6 | 0.7 | 39.4 |
| Urine treated straw + soyabean (urease) | 58.6 | 56.2 | 57.9 | 0.7 | 40.3 |
| Urea treated straw | 46.9 | 46.8 | 53.4 | 2.3 | 39.4 |
| Urea treated straw + soyabean | 49.3 | 50.0 | 57.2 | 2.5 | 36.5 |
| Anhydrous-$NH_3$, stack | 56.1 | 56.7 | 65.7 | 3.1 | 33.1 |
| Aqueōs $NH_3$, stack | 58.0 | 61.6 | 57.2 | 1.7 | 31.2 |
| Anhydrous $NH_3$, vacuum, stack | 57.1 | 57.6 | 64.6 | 2.4 | 32.2 |
| Anydrous $NH_3$, oven | 55.9 | 59.7 | 60.9 | 1.5 | 32.2 |
| Beckmann (NaOH) treated straw | 69.2 | 78.1 | 72.8 | 2.6 | 21.1 |
| Wet (NaOH) treated straw (LCM) | 69.0 | 74.9 | 72.0 | 1.7 | 25.4 |
| Dip (NaOH) treated straw | 73.3 | 88.4 | 74.8 | 4.8 | 23.0 |
| Dip (NaOH) treated straw + urea | 65.5 | 86.6 | 75.2 | 6.1 | 25.0 |
| Dry (NaOH) treated straw | 65.9 | 67.3 | 67.8 | 0.6 | 28.8 |
| Dry (NaOH) treated straw, pelleted | 65.5 | 67.4 | 64.2 | 0.9 | 30.8 |
| Method means | 59.4 | 63.3 | 63.3 | 1.10 | – |

The results from the in vitro and in sacco determination compare fairly well with the in vivo results. There was a tendency for underestimation of straw with low digestibility with the in vitro method whereas the in sacco method seemed to underestimate the digestibility of straw with a high in vivo digestibility. The reason for this might be that the rate of passage through the forestomachs is lower for untreated straw and higher for wet-treated straw relative to the 48 h incubation in vitro and in sacco.

Solubilisation of indigestible substances in vitro and in sacco would have the same effect. In any case, the potential for improving the quality for various materials should be given some consideration before embarking on a comprehensive research programme. A screening by use of the in sacco or in vitro technique could be a useful tool here.

### Accessibility and Price of Chemicals and Equipment

One of the most obvious prerequisites when planning chemical treatment of straw is that the chemical must be available at an acceptable price. If some special equipment like plastic tubes, plastic sheets, cans, pressure tanks etc. are needed this should also be available and not too expensive. As far as price of chemicals is concerned, it is difficult to generalise. Some countries produce their chemicals locally and can sell them at a moderate price whereas others have to import the chemicals and they make the price high to reduce currency outflow.

Prices are also changeable; high prices may be due to small sales volume. If a market is developed and a distribution network is built up the price may drop dramatically. Therefore there is no reason why these factors should be looked upon as constants.

### Adverse Effects of Chemicals

One question that has to be asked when planning a

research project on chemical treatment of fibrous by-products is whether or not the chemical in question has any adverse effect on the animals or the environment. Large quantities of alkali are thought to have undesirable effects upon the health of the animals consuming these feeds. Long term experiments in Denmark over at least 3 lactations with 40% dry (NaOH) treated straw in the diet have not revealed clinical sign of sickness among the animals.

In Norway, individual cows have consumed 700 - 800 g of NaOH in Dip treated straw for short periods without showing any indications of health problems. One may conclude from the evidence available at present that the knowledge about alkali (NaOH) tolerance in farm animals is somewhat limited. Further studies at this point are therefore desirable. Ammonia used for straw treatment is not considered to represent any health problems under normal circumstances whereas the risk of urea poisoning is well known and should be recognised.

In Norway, NaOH treatment of straw according to the Beckmann procedure has been practiced for more than 40 years. By this method the excess NaOH and some organic matter from the straw is washed into rivers and lakes. This has caused a pollution problem which is not accepted any longer from an environmental point of view. "Closed methods" have therefore been developed. By these methods all the chemical (NaOH) absorbed by the straw passes through the animals and is excreted through the urine and faeces. If the yearly precipitation is high, this is not considered to cause any soil pollution. In arid areas, this is a point that deserves special attention.

There is no doubt about the technical feasibility of many of the methods developed over the last decades. An economic assessment of the methods is therefore pertinent. To obtain a certain degree of precision, this assessment must be done locally because the conditions differ so widely. One rule of thumb might be that the treatment is economical if the cost of treating 100 kg of straw is less than the value of the energy gained by the treatment. With the ammonia or urea treatment the "protein" value of the chemical should be added.

To encourage farmers to use fibrous by-products as a feed resource, and to save foreign currency, some countries offer financial support to those who treat straw chemically. In some countries such as Bangladesh the price of chemical is subsidised by the Government. In other countries the support may be given to the farmer based on the quantity of straw treated. If imported cereals for animal consumption is subsidised, there is no reason why support should not be given for upgrading of by-products for feeding.

**Research Programme**

Countries in which straw and other crop residues constitute the major part of the diet for the farm animals should give priority to research on utilisation of these feed resources. A research programme in this area may be organised in different ways. One approach is the proposal put forward by Jackson in his excellent report. In spite of the new developments that have taken place since 1978 many elements of his proposal are still relevant. The research programme could be divided into three projects which should be closely linked to each other.

- Basic research
- Applied research
- Field research

One person could be in charge of each of these three projects and one could serve as a coordinator. The three or four persons should work as a group and they should meet frequently to discuss problems and administrative matters. The importance of getting hold of well-qualified persons for these jobs should be emphasised.

### Basic Research

Basic research may be needed to study a great number of factors that may influence the effect of a treatment: e.g. dosage of chemical, time of treatment, moisture content of the material etc. Laboratory research is also useful for screening purposes in which a number of materials may be tested as to their initial value and their response to chemical treatment. Fundamental studies on rumen fermentation, rate of passage and measurement of digestibility may be useful to enable the project personnel to draw general conclusions and to solve problems that cannot be studied in a field situation. These type of studies may well be undertaken at universities and some experimental stations. Often such research is suitable for post-graduate students and visiting scientists. Findings from such experiments are important also for the teaching of other students.

### Applied Research

Feeding trials at the production level are necessary to prove the value of a feedstuff in a real situation. Such experiments should be carried out at university farms or experimental stations with a sufficient number of animals to enable the researcher to show significant differences statistically. Young, growing animals may be used for this purpose and 10 to 15 animals per treatment would usually be required. The animals should preferably be fed individually and the feed should be weighed every day.

The animals should be weighed regularly. If the animals are slaughtered at the end of the experiment the carcass quality should be judged. If lactating animals are used the milk production should be recorded at least 4 to 5 days per week. The animals should normally be fed ad libitum. If restricted feeding is practiced, periods of at least 2 - 3 weeks of appetite feeding should be included to allow estimations of feed intake. An economic appraisal of the results should also be carried out.

Efforts should be made to balance the diet with regard to protein, minerals and vitamins. Since concentrates are scarce in most developing countries, by-products from the milling industry, fishing industry and slaughter houses are often the most common feed supplements. Small amounts of fresh grass, weeds, water plants, Leucaena and others have also proved to be valuable supplements. The potential for NPN utilisation should also be explored in such situations. Treatment of crop residues with urea/ammonia has the advantage over NaOH in that it increases the N-content of the material by 0.5–1.5%. This nitrogen is undoubtedly used to some extent by the rumen microbes for protein synthesis.

On top of this, small supplements of undegradable protein may improve the performance of the animals considerably. The role of supplementation for an optimal utilisation of diets based on fibrous residues and by-products is discussed in depth by Preston and Leng. There has been a tendency for university teachers to consider themselves very academic in that it is not their role to do a practical job in the field. In applied science it is absolutely necessary that the teacher throw off the white coat and work with his hands.

#### *Field Research*

Since much technical knowledge already exists, work should start at the field level as soon as possible. It is not so necessary to introduce a package of technology, but rather to become familiar with the local conditions and identify the problems that may arise when attempting to introduce a new technology. Under such a scheme three experimental diets could be tested.

- Traditional feeding.
- Diet based on untreated straw + urea and minerals.
- Diet based on treated straw + urea and minerals.

Other supplements may also be of interest depending upon the local conditions. Experiments should be carried out at agricultural schools or at private farms with sufficient numbers of animals. The farmers chosen should be clever and motivated to cooperate. If possible the hosts for field experiments should be gathered at the beginning of the project to be introduced to the research plans and the actual methods for treatment. The possibility for cooperative treatment of crop residues, for instance on a village basis, should be explored.

Crucial for a successful project is that the person in charge of the field research has a solid theoretical background and practical experience. He/she should have practiced straw treatment elsewhere or worked for some time under a person well qualified in the field. The project leader should have sufficient funds to buy the chemicals and small items needed for the research work. Funds will also be necessary for transport and technical assistance.

The field research will include treatment of the material under the control of the project leader, planning of the trials and the actual research work. Occasional weighing of the rations may provide some information about the intakes of straw. Weighing of animals is often a problem in the field, but chest measurements may give an indication of growth rates and body development. Milk production is, of course, easy to measure in lactating animals.

Samples from the feeding trials may be taken to the experimental station or the university for simple chemical analyses to characterise the quality of the untreated and the treated material. An economic evaluation of the treatment is desirable as far as this is possible under primitive conditions. If the wanted chemical is not available locally, the project leader should investigate how to get hold of it. If it is available, but at a very high price, the project leader should purchase a greater quantity which normally results in a price reduction.

A number of other problems may also arise, problems that may be difficult to foresee. However, frustration will not lead to any good. It is better to face the problems, and try to solve them trusting that a successful use of crop residues and agro-industrial by-products as feed sometime in the future will mean a better life for millions of people throughout the world.

The leaders of the field, applied and basic research should form a group that meet frequently to discuss common problems, the progress made and the future of the research programme. As part of the procedures for research into the topic training courses may be arranged for technical personnel at each centre. Some key persons may even go abroad for technical training: not to learn a technology which could be transferred directly, but rather to learn some basic concept with regard to treatment of fibrous by-products.

International seminars and workshops such as those organised recently in Bangladesh, Sri Lanka, Tanzania, Kenya, Egypt and elsewhere are stimulating for the participants who get an opportunity to discuss common problems and exchange information with colleagues with similar interest. It is also important that treatment of fibrous by-products is discussed in relation to various systems for animal production to place it in a broader context. In spite of the many seminars organised, there seems to be a lack of information between the various groups working in this field. Therefore the proposal came up at the workshop in Alexandria last year that an information officer be employed for a period of 3 to 5 years who could disseminate information among the various groups.

### Evaluation of Crop Residues and Agro-Industrial By-products Using the Nylon Bag Method

The use of nylon bags to study rumen fermentation of feeds is not new. The technique has been used for a long time but mainly for definite incubation periods as an estimate of in vivo digestibility. In 1975 it was felt that measuring degradability of proteins by post ruminal collection was not appropriate and attempts were made to adapt the nylon bag method to describe the rate and the extent of degradation of protein supplements. In other words, the nylon bag was used not only to give information as to the possible extent of digestion or degradation but also the events that led up to it, i.e. the rate at which degradation occurred. This measurement of rate is particularly important because it provides information on factors affecting intake of roughages.

Voluntary intake of roughages in ruminants is not only affected by the digestibility, but also by the rate at which digestion occurs because both factors influence turnover rate. In many respects, therefore, a description of feeds using the nylon bag is much more instructive and gives more information than the description of in vivo digestibility. The nylon bag method has the advantage of simplicity. It requires only some nylon bags, rumen cannulated sheep, goats or cows and an oven to dry samples. Equipped with these tools, a lot of information can be obtained without further sophistication.

The nylon bag incubations have advantages in so far as digestion can be studied in the environment to which it is applicable; that is the environment of in vivo rumen fermentation. Studies made during in vivo fermentation have the added advantage that there is no problem of temperature control, there is no problem of removal of end products and so on; problems which have to be coped with as soon as rumen fluid is taken into a laboratory for in vitro fermentation. It must be pointed out, however, that in the method recently developed by Czerkawski and Breckenridge, continuous fermentation may sometimes be adapted for routine laboratory use.

Some researchers done a study to evaluate the crop residues and agro-industrial by-products using the nylon bag method. The description of these study is briefly described below.

### Description of Technique

Nylon bags with a mesh size of 20–40 um are most suitable. Smaller mesh sizes can be used, in fact the mesh size is a compromise. It must allow entry of rumen microbes and escape of accumulated gas on the one hand, and on the other hand it must keep losses of solid particles to a minimum. Accumulation of gas in the nylon bag can cause the bags to float on top of solid digesta. The size of the bags they used is about 140 × 90 mm and they use 3–5 g of dry samples. It is possible to use larger bags and larger samples and in fact the size of the sample is quite often dependent upon the amount of residue required for analysis.

It is essential and extremely important that the materials incubated are able to move freely within the bag, so as to avoid the formation of micro-environments in the bag with resultant poor replications. The diameter of the rumen cannula is also sometimes a function of the size of the bags. For cattle, larger cannulas and more bags can be used if required.

The preparation of samples must, as far as possible, represent materials as they appear in the rumen had they been consumed naturally by the animal. This is not possible however as this would require use of animals with oesophageal cannulas. For green materials, succulent materials and silage a mincer is more appropriate with a 5 mm screen size.

Many workers recommend that bags are anchored with a 25 cm of nylon cord at the top of the cannula in sheep and about 50 cm in the rumen of cattle. This length allows the bags to move freely within digesta both in liquid and solid phase. Some workers use a weight attached to the bags to ensure that the bags do not float on top of the rumen contents while others find that additional weights give little or no improvement in accuracy. Recently, we have used a method of anchoring the bags to a nylon tube which is illustrated in Figure 1. This nylon tube can hold about 5 bags at the same time. The system simplifies the withdrawal of the bags since individual bags, on an individual cord, can become tangled and difficult to withdraw from the rumen.

The most appropriate times at which bags can be withdrawn from the rumen in order to best describe the disappearance rate, depend on the shape of the degradation curve. It is not possible to be absolutely specific in the guidance one can give on time for withdrawal of bags. For many protein supplements, withdrawal of bags at 2, 6, 12, 24 and 36 h gives an adequate description. For hay, straw and other fibrous materials, longer incubation intervals are required. In this case, bags are taken out at 12, 24, 48 and 72 h.

The most important source of variation is the between animal component. As a result it is necessary to use more than one animal to get an accurate estimate of a particular feed. Quite often some researchers find that 3 animals is a good number to work with. They find, for instance, that between sheep, variation of dry matter disappearance was 6.2% while the variation between days was 4.9%; and the variability between bags only 3.3%. From this information it is possible to provide guidance as to the importance of replication between bags, between days and between sheep. Table 3 is taken from Mehrez and Orskov.

*Table 3. Estimated variance of the mean dry matter disappearance using the rumen bag technique for various numbers of bags (b), days (d) and sheep (s)*

| *Replicates* | | | *n* | *Variance of mean* |
|---|---|---|---|---|
| Bags | Days | Sheep | (b × d ×s) | (% of mean) |
| 1 | 2 | 3 | 6 | 3.43 |
| 2 | 1 | 3 | 6 | 4.25 |
| 4 | 1 | 2 | 8 | 5.96 |
| 2 | 2 | 2 | 8 | 4.74 |
| 1 | 2 | 4 | 8 | 3.19 |
| 4 | 2 | 2 | 16 | 4.53 |

Few comparisons have been made but results so far suggest that if sheep or cattle are given the same feeds then there is little or no difference in degradation rates measured with nylon bags. The use of cattle has the advantage that a larger number of bags can be used. On the other hand sheep are much cheaper to keep and use. It is more important to remember that the feeding level is very important; this sometimes dictates the particular species that has to be used.

### Effect of Diet on Rate of Degradation

The diets can have a great effect on rates of degradation, both of fibrous feeds and of protein supplements. Indeed, this is one of the aspects of digestion that is very interesting and important to study; namely, the effect of the rumen environment on degradation of dietary constituents. It is therefore important that protein degradabilities are determined in the rumen of an animal receiving a diet similar to that which is used in practice and for which the results of protein degradability will have to be applied. It has been shown for instance that protein supplements of vegetable origin are normally degraded more slowly in the rumen of animals given high concentrate diets compared with high roughage diets.

In order to exploit fully the information provided by the description of degradation, whether it be degradation of protein supplements or degradation of fibre, it is essential that the incubation times chosen are such that both the sensitive part of the curve and the asymptote are adequately described. It may well be that for some supplements it is necessary to obtain more incubation times or to correct the incubation times used the first time because of lack of experience of a particular type of feed. Having ascertained that the curve is adequately described it is then possible to apply some mathematics to it in order to assist the interpretation. The researchers have chosen an exponential description of degradation which fits most situations. The expression is of the type: $p = a + b\ (1 - e^{-ct})$.

The reason why this equation is very suitable is due to the fact that the constants in the exponential equation have some meaning. First of all 'p' is the degradation which has taken place during the time 't'; 'a' is essentially similar to the very rapidly disappearing fraction because it is the intercept and highly correlated with the water soluble fraction; 'b' is that portion of the feed which will be degraded in time; 'c' is the degradation rate of the 'b' fraction. It follows that 'a' + 'b' give some measure of the potential digestibility or the potential degradability of the feed being tested.

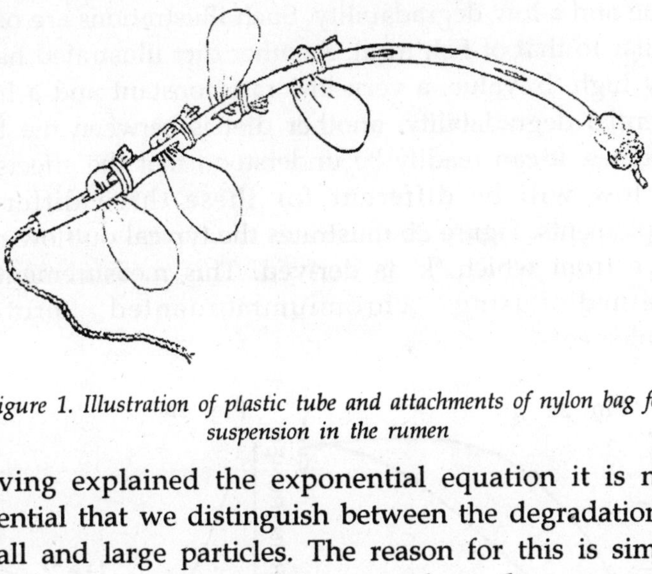

*Figure 1. Illustration of plastic tube and attachments of nylon bag for suspension in the rumen*

Having explained the exponential equation it is now essential that we distinguish between the degradation of small and large particles. The reason for this is simply that small particles can flow out from the rumen and enter the lower gut without further reduction in particle size. As a result we can now apply another equation—namely, the rate of outflow of small particles. It will then be logical to assume that the effective degradability of small particles will depend on solubility, i.e. on the 'a' value and on the rate at which the 'b' fraction is degraded. It will also depend on the rate of outflow. We have managed to put together the two expressions in one which we call the effective degradability which is equal to:

$$P = a + \frac{bc}{c + k}$$

Now, it is possible to examine the extent to which the effective degradability will vary at different rates of outflow. Figure 2a illustrates three different protein supplements which show a different pattern of degradation. One has a high 'a' value, very little of a 'b' value and a low degradability. Such illustrations are often similar to that of fish meal. Another diet illustrated has a very high 'b' value, a very low rate constant and a high potential degradability; another diet is between the two extremes. It can readily be understood that the effects of outflow will be different for these three different supplements. Figure 2b illustrates the typical outflow rate curve from which 'k' is derived. This measurement is obtained using chromiummordanted protein supplements.

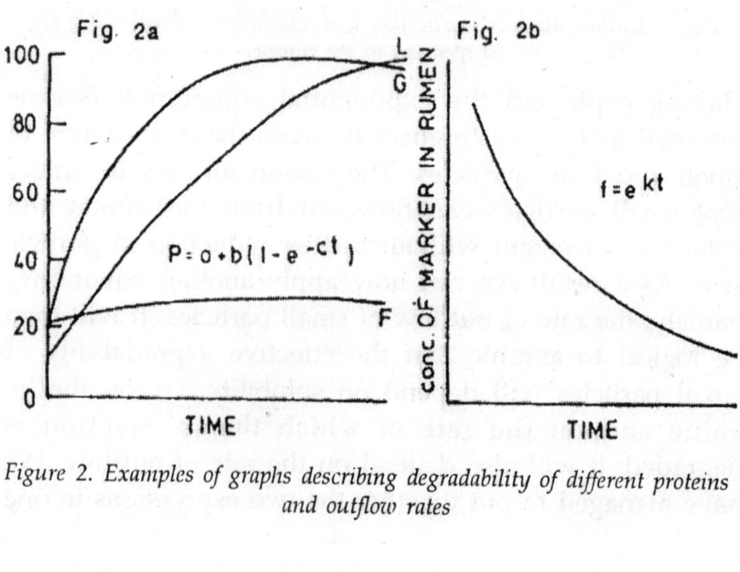


*Figure 2. Examples of graphs describing degradability of different proteins and outflow rates*

The rate of outflow will have very little effect on such products as fish meal since the 'b' value is very low and the effective degradability will be very similar to 'a' (Figure 3).

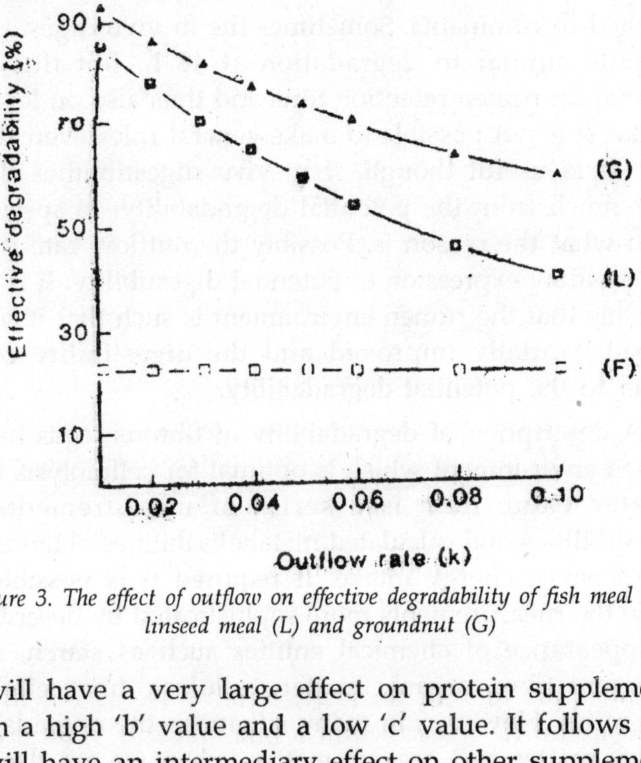


*Figure 3. The effect of outflow on effective degradability of fish meal (F), linseed meal (L) and groundnut (G)*

It will have a very large effect on protein supplements with a high 'b' value and a low 'c' value. It follows that it will have an intermediary effect on other supplements which have a high 'c' value combined with a high 'b' value. It is important that this is understood because it is not possible to use a constant correction or to use a particular incubation time to represent effective degradability.

### Fibrous Feeds Degradation

The same expression as mentioned before can be used to

describe the degradation of fibrous feeds except that the outflow rate of small particles cannot directly be applied. The (a + b) value will give some expression for potential digestibility for the fibrous feeds.

The potential degradability is however seldom reached in ruminants. Sometimes the in vivo digestibility is quite similar to degradation at 48 h. But this will depend on rumen retention time and thus also on level of intake. It is not possible to make general rules even about that. It is useful though, if in vivo digestibilities differ very much from the potential degradability, to speculate as to what the reason is. Possibly the outflow rate is too fast to allow expression of potential digestibility. It is also possible that the rumen environment is such that it could be substantially improved and the digestibility come closer to the potential degradability.

A description of degradability of fibrous feeds in the rumen environment which is optimal for cellulolysis is of greater value than is a series of measurements of digestibilities and calculated metabolisabilities obtained at maintenance energy intake. If required it is possible to make the measurements more sophisticated by describing disappearance of chemical entities such as starch, acid detergent fibre, organic matter, cellulose, hemicellulose and so on. However, in many situations the degradation of dry matter and organic matter, when we are dealing with fibrous feeds, is of the greatest value.

#### *Calculation of Rate of Degradation*

Description of the disappearance of protein or fibre by the exponential equation is easy to do with a computer or a "scientific" calculator; this permits the estimation of the residual standard deviation of the equation. However, if no computer facilities are available it is quite possible to caluculate the equation by simple arithmetic. Essentially, it is necessary to describe the curve as in Figure 4. The procedure is to fit the curve to the measurements obtained. It can be seen that the intercept 'a' is '6', the asymptote in the equation here is found to be '92', i.e. (a + b) is 92% which means that 'b' is '86'.

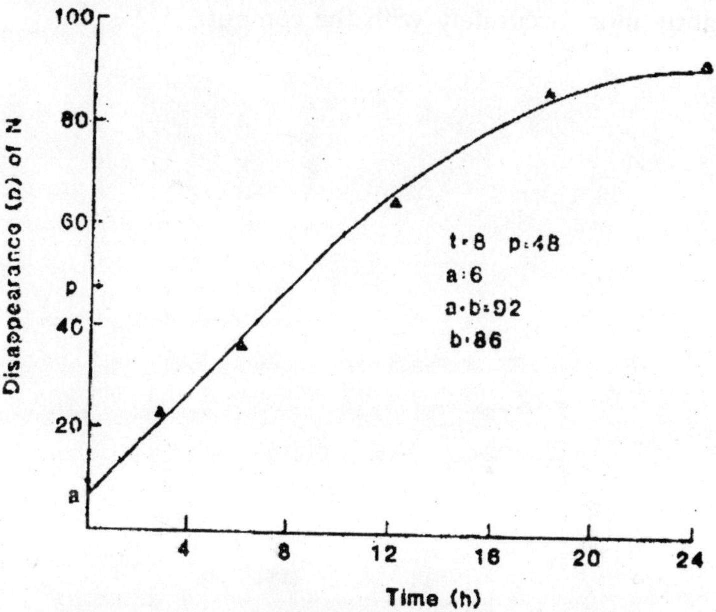


*Figure 4. Illustration of calculation of rate of degradation from disappearance curve fitted by eye*

If we now take a value on the curve where degradation is occurring most rapidly we find that 'p' is '48'.

It is now possible to describe the equation as:

$$e^{-ct} = (a + b_{b} - p)$$

which means that:

$$e^{-c3} = (6 + 86_{86} - 48) = 0.5116$$

By taking the natural log on both sides of the equation, it is found that:

$$c = 0.084$$

All the constants in the equation are now known and they are in agreement with those which can be obtained slightly more accurately with the computer.

# 11

# Organic Crop Production

Organic agriculture is an ecological production management system that promotes and enhances biodiversity, biological cycles and soil biological activity. It is based on minimal use of off-farm inputs and on management practices that restore, maintain and enhance ecological harmony. As organic farming and marketing entered the 1970s, it began to develop as an industry. As a result, a clearer definition was needed to distinguish it and its products from conventional agriculture. This was no straightforward task.

Environmental issues and other alternative agriculture philosophies had created diverse notions about what organic agriculture was and what it should be. A particularly problematic image grew unexpectedly from the anti-pesticide movement of the 1960s. This was the romantic notion that organic simply meant "doing next-to-nothing." In this exploitative approach, not only were pesticides avoided, sound farming practices that built the soil were also largely ignored.

The results achieved on such farms were predictable, as yields were low and the quality poor. These approaches became collectively known as organic by neglect and are a far cry from the responsible farming models proposed by Albert Howard and J.I. Rodale. It is unclear how many farmers actually chose to farm "by neglect" and advertise themselves as organic over the years. However, this extreme representation of organic agriculture was quickly taken up by critics who tried to characterise all of organic agriculture as soil depleting and unproductive.

To counter this, current standards for certified organic production require an "organic plan" outlining the use of soil building activities and natural pest management. There is a further notion that organic farming also describes farm systems based on soil building, but that continue to use some prohibited fertilisers and pesticides in a limited or selective manner. A USDA study of U.S. organic farms made note of many such individuals who readily and sincerely referred to themselves as organic farmers. While these growers were largely conscientious and would, in most instances, fall under the modern umbrella of "sustainable farmers," industry standards evolved to preclude all synthetic pesticides or commercial fertilisers.

Organic systems are generally more labor intensive. Studies done in the late 1970s by Washington University, for example, found that about 11% more labor was required per unit of production where agronomic crops were concerned. This difference can be much greater where horticultural crops are involved, and farm size may be limited accordingly. However, technological innovations in organic horticultural production are helping to narrow the gap. Organic systems are also more information intensive, requiring additional management time in planning, pest scouting, and related activities. For this reason, organic management can be better done if a farm is not too large. Essentially, the notion that organic systems are only possible on very small farms is a false one. Given the range of acceptable technologies available, organic agriculture can be sized to fit a wide range of farms and enterprises.

Throughout its early history, organic agriculture was treated with either hostility or apathy by the USDA, land grant universities, and conventional agriculture in general. Fortunately, the atmosphere for discussing and investigating organics has improved considerably. While it did not become boldly evident until the 1990s, the tide actually began turning in the late 1970s and early 1980s. A number of factors precipitated this change, among them the growth in the organic industry. Serious money demands serious attention. Also critical, from the perspective of the research community especially, were some landmark studies that lend credibility to organic farming as a truly viable option for American agriculture.

In 2002, when the USDA adopted the National Organic Standard that spells out what farmers and food processors must (and must not) do to be certified "organic," the organic industry already had a long history of relying on third-party certifiers to ensure the integrity of their products and practices. Under this system, a state-run or accredited private agency (the third-party) evaluates farmers and processors to see whether they conform to the standards of the National Organic Program (NOP). Those who do can then market their products as "USDA Certified Organic" and display the official USDA organic seal on their packaging.

In essence, certification is largely about integrity—assuring that the buyer is getting what he or she is paying for. Certified organic production, then, means production by approved organic methods, with additional pains taken to eliminate contamination with prohibited materials and commingling with conventional products. There is a common misconception that certified organic means "pesticide residue-free." Consumers have a right to expect little or no pesticide residue on certified organic crops because none are used in their production. However, ours is a dirty world in which pesticides and their break-down products are ubiquitous. This is only to be expected in a national farm system where more than 99% of all applied farm chemicals miss the target organism.

### Principles of Organic Farming

There are several compelling principles that characterise certified organic farming. They include biodiversity, integration, sustainability, natural plant nutrition, natural pest management, and integrity. Most organic operations will reflect all of these to a greater or lesser degree. Since each farm is a distinct entity, there is a large degree of variation.

As a general rule, diverse ecosystems in nature have a higher degree of stability than those with only a few species. The same is essentially true for agroecosystems. Farms with a diverse mix of crops have a better chance of supporting beneficial organisms that assist in pollination and pest management. Diversity above ground also suggests diversity in the soil, providing better nutrient cycling, disease suppression, tilth, and nitrogen fixation. Good organic farmers mimic the biodiversity of nature through practices like intercropping, companion planting, establishment of beneficial habitats, and crop rotation (sometimes referred to as companion planting across time). The effort to increase biodiversity works hand-in-hand with enterprise diversity, which is often (but not necessarily) an objective on organic farms.

The drive to build biodiversity in organic systems encourages diversity among enterprises, but not as isolated or independent entities. Good organic operations integrate their various enterprises. A good example can be seen among Midwestern organic crop and livestock operations. The typical Midwestern organic operation ties the needs of crops and livestock together in a practical and elegant way. The forage and grain needs of ruminant livestock make for a diverse mix of crops. Particularly valuable is the inclusion of legume forages for ruminant feed.

Forage legumes in rotation fix a sustainable supply of nitrogen in the soil that feeds subsequent non-legume crops in rotation. Manure from the livestock enterprises is conserved as a nutrient resource and recycled back to the crop fields. Farms such as these have the additional advantage of greater economic sustainability, as their risks are spread over several livestock and crop enterprises.

In addition to the greater economic sustainability afforded by enterprise diversification, organic farmers are often able to reap market premiums for certified production. However, since many organic enterprises realise somewhat lower marketable yields, this has not always translated into higher profits or greater economic sustainability. As more and more organic growers enter the marketplace, it is likely that premiums will stabilise at modest levels and may vanish for some crops. Organic producers need to look well ahead and be aware of shifting trends.

As alluded to earlier, many U.S. organic farms perform well on many of the measurable indicators associated with sustainability, such as energy consumption and environmental protection. However, sustainability is an ideal, and the best that can be said is that current organic farms are closer to the ideal than most alternatives—certainly closer than comparable conventional farming operations. The extent to which traditional organic agriculture philosophy influences the adoption of sustainable practices has only been touched upon.

### Plant Nutrition

Even though we require the same basic "stuff" to live, it is somewhat challenging to draw simple comparisons between the nutritional needs and processes of plants and those of animals. Plants are able to photosynthesise to make sugars, which are ultimately synthesised into proteins and other plant constituents. Humans and other animals, by contrast, can obtain energy foods, proteins, and vitamins only by consuming plants or other animals. Both plants and animals also require minerals.

Humans and other animals extract minerals, along with sugars and proteins, from the food they eat. Plants, too, obtain minerals—and a wide range of vitamins, antibiotics, and other useful compounds—through digestion. However, plant digestive systems are not internalised as they are in animals. Plants must rely on the external digestive processes of the soil system within reach of their roots—a zone called the rhizosphere.

Critics are often under the illusion that organic farmers believe plants obtain all their nutrients from an organically managed soil in a chemically organic form. While a few organicists may believe that, the majority recognises that digestion processes in the soil release minerals in forms similar to those applied as commercial fertilisers. Unfortunately, the notion that organic farmers are naïve and ignorant about basic agronomy is a red herring that has often foiled intelligent discussion about the pros and cons of the system.

Among the facts that are often obscured is that plants can and do absorb significant amounts of large organic molecules from the soil; herbicide and systemic insecticides are among these. In healthy soils they also absorb vitamins, chelated minerals, hormones, and other beneficial compounds. The organic philosophy of crop nutrition begins with proper care and nourishment of the organisms responsible for the soil digestive process.

Organic farmers believe this is best accomplished by avoiding toxic chemicals and practices—like excessive tillage—that are harmful to soil organisms, as well as by the addition of organic matter and natural rock minerals. Conventional systems, in contrast, try to circumvent the soil's digestive process and provide needed minerals to the plant directly, in a soluble form. From the organic perspective, the conventional approach has several flaws.

- Applying large quantities of soluble fertiliser to a crop only one, two, or three times per season floods the plant with those nutrients, causing nutritional imbalances that lead to crop diseases, insect infestations, and reduced food quality.
- Failure to support and care for soil biotic life, along with other practices that are downright destructive, ultimately leads to its decline. As a result, plants lose out on the vitamins and other beneficial products these organisms produce, tilth is reduced, and the soil becomes increasingly dependent on synthetic inputs.
- Conventional fertilisation tends to concentrate on a limited number of macronutrients, even though the need for at least 13 soil minerals is scientifically recognised. This skewed focus is also responsible for generating imbalances in the plant.

- Application of large amounts of soluble nutrients can stimulate certain problem weed species.
- Soluble nutrients—especially nitrate—are prone to leaching, which can cause a number of environmental and health problems.It is organic farming's approach to soil building and plant fertilisation that is the true basis for the belief that organic food and feed has superior nutritional value, much more so than the absence of pesticide residues, which has drawn the spotlight ever since the 1960s.

**Pest Management**

Whether conventional or organic, all farmers are concerned with pests. They spend a lot of time and resources controlling them. However, in the organic "world view," pests—whether weeds, insects or diseases—are not simply scourges. They are indicators of how far a production system has strayed from the natural ecosystems it should imitate. Certain weeds, for example, tend to predominate when soils are too acidic or too basic; some become a problem when soil structure is poor and conditions become anaerobic; others may be stimulated by excessive fertiliser or manure salts.

Organic proponents also believe that insect pests are attracted to inferior or weak plants—the result of poor crop nutrition. Their logic continues by asserting that pests are naturally repelled by vigorous, well-nourished plants. This belief is often challenged, and significant research remains to be done. As scientific understanding has grown, insect pest outbreaks are also being understood as imbalances in the whole agroecosystem and how it is managed.

In nature, massive pest outbreaks are relatively rare and short-lived, due to the presence of natural predators, parasites, and disease agents that quickly knock the pest numbers back down to a moderate level. In farming systems that inadvertently destroy or otherwise fail to support the natural control complex, pest problems are routine and, typically, worsen with time. The farmer becomes increasingly addicted to costly and extreme control methods to produce a crop. Most organic growers consider pesticides to be a cause of agroecosystem imbalances and employ allowed natural pesticides as little as possible.

### Integrity

Integrity refers to the systems in place and actions undertaken to assure that consumers of organic products get what they pay for. Consumers have a right to expect that the organic food they buy not only be raised by organic methods but be protected from contamination and from commingling with non-organic products. While the responsibility for much of this rests with others in the organic marketing chain, many certified organic growers need to incorporate additional practices that work to assure the integrity of their products.

Proper record keeping is very important in this regard, though growers are often reluctant to spend much time on it. Among the more important production practices in the field are buffer strips, which reduce chemical drift from neighboring fields and roadsides, while also serving water and soil conservation objectives.

### Tools and Practices of Organic Agriculture

The tools and practices of organic agriculture include traditional alternatives—crop rotation, manuring, liming, etc.—long recognised as important to a sound production system. They also include more contemporary practices and materials that research and keen observation over time have contributed. The following list of tools and practices is not intended to be comprehensive, though the primary options are addressed. Note, too, that each farm operation will employ its own combination of tools and practices to build a working organic system.

#### *Planned Crop Rotation*

Essentially a tool for annual cropping systems, crop rotation refers to the sequence of crops and cover crops grown on a specific field. Particular sequences confer particular benefits to long and short-term soil fertility, and to pest management. Agronomic operations are especially dependent on crop rotations that include forage legumes. These provide the vast majority of the nitrogen required by subsequent crops like corn, which is a heavy consumer of that nutrient. Even when livestock are present to generate manure, the animals are largely recycling the nitrogen originally fixed by legumes in the system. An example of a basic agronomic rotation, typical of that found on midwestern organic farms, is shown in Figure 1.

The basic Midwestern rotation demonstrates the elegant way in which a whole farm system can function:

- Legumes fix nitrogen in the soil, providing for subsequent non-legumes in the rotation.
- Several insect pest cycles are interrupted, especially those of the northern and western rootworm species that can be devastating to corn.
- Several plant diseases are suppressed, in cluding soybean cyst nematode.

- Weed control is enhanced as perennial weeds are destroyed through cultivation of annual grains; most annual weeds are smothered or eliminated by mowing when alfalfa is in production.
- Livestock manures (if available) are applied just in advance of corn, a heavy nitrogen consumer.
- All crops can be marketed either as is or fed to livestock to be converted into value- added milk, meat or other livestock products.

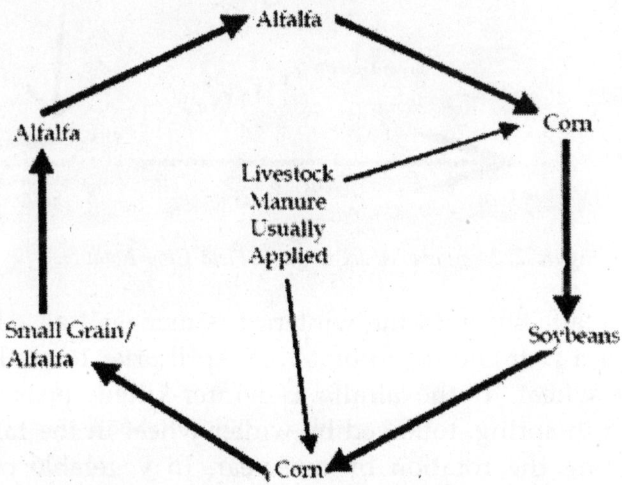


*Figure 1. Corm Belt Model Organic Field Crop Rotation*

Another example of an agronomic crop rotation – this one suitable to drier, western climates – is typified by the Quinn Farm in North-Central Montana and presented in Figure 2. Bob and Ann Quinn's rotation begins with the most reliable cash crop, hard red winter wheat, fall-seeded after alfalfa. Weeds are controlled following harvest and the land reseeded to lentils, kamut, or durum wheat the following spring. Switching from a winter grain to a spring grain helps to break weed cycles and optimises soil moisture. In the next year, another spring grain or buckwheat is planted and undersown with alfalfa.

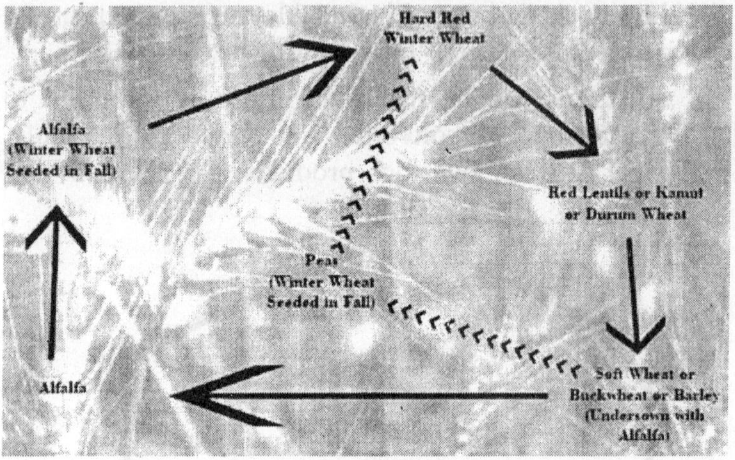


*Figure 2. Montana Model Organic Field Crop Rotation*

If the alfalfa survives the winter, it is managed as a hay crop for a year and incorporated in April prior to seeding winter wheat. If the alfalfa is winter-killed, peas are planted in spring, followed by winter wheat in the fall—shortening the rotation by one year. In vegetable crop rotations, nitrogen fixation and carry-over is also important, though it plays second fiddle to pest management. The well-known market gardener Eliot Coleman recommends an eight-year rotation, as shown in Figure 3.

The rationale for Coleman's eight-year rotation follows. Since he gardens in the Northeast, some of the details reflect those constraints. Potatoes follow sweet corn...because research has shown corn to be one of the preceding crops that most benefit the yield of potatoes. Sweet Corn follows the cabbage family because, in contrast to many other crops, corn shows no yield decline when following a crop of brassicas. Secondly, the cabbage family can be undersown to a leguminous green manure which, when turned under the following spring, provides the most ideal growing conditions for sweet corn.

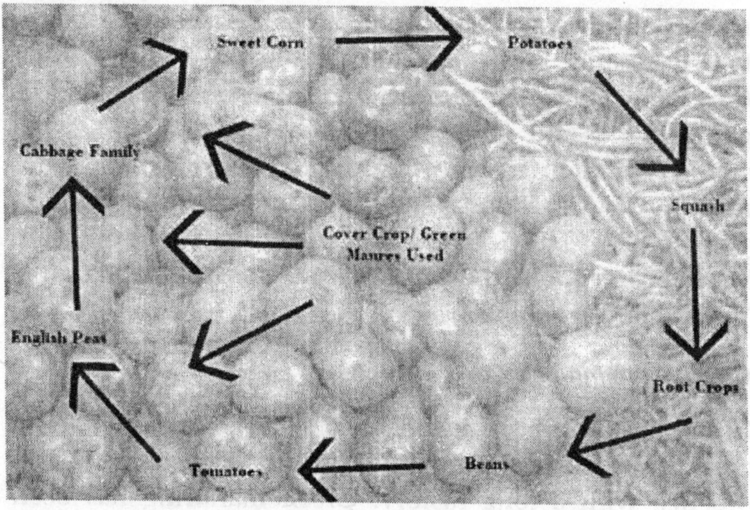


*Figure 3. Ellet Coleman's Vegetable Crop Rotation*

The Cabbage Family follows peas because the pea crop is finished and the ground is cleared [early] allowing a vigorous green manure crop to be established. Peas follow tomatoes because they need an early seed bed, and tomatoes can be undersown to a nonwinter-hardy green manure crop that provides soil protection over winter with no decomposition and regrowth problems in the spring. Tomatoes follow beans in the rotation because this places them 4 years away from their close cousin, the potato. Beans follow root crops because they are not known to be subject to the detrimental effect that certain root crops such as carrots and beets may exert in the following year.

Root Crops follow squash (and potatoes) because those two are good "cleaning" crops (they can be kept weed-free relatively easily), thus there are fewer weeds to contend with in the root crops, which are among the most diffi cult to keep cleanly cultivated. Second, squash has been shown to be a beneficial preceding crop for roots. Squash is grown after potatoes in order to have the two "cleaning" crops back to back prior to the root crops, thus reducing weed problems in the root crops. Georgia growers Ed and Ginger Kogelschatz use a somewhat simpler rotation scheme that divides most garden crops into four basic classes that are then sequenced for a 4-year cycle. They have adapted this concept from Shepherd Ogden, the author of Step By Step Organic Vegetable Gardening.

#### *Green Manuring and Cover Cropping*

Green manuring consists of incorporating into the soil a crop grown for the purposes of soil improvement. It is a practice with a long history. Green manuring has been ignored in recent years as a serious option for soil improvement because the traditional practice entailed planting a full-season cover crop. This removed the field from commercial production for a whole season. Interest has returned, however, since green manuring strategies have been combined with cover cropping schemes. Cover cropping is growing a crop for the purpose of soil and nutrient conservation.

It is a more contemporary concept than green manuring, in crop agriculture. The two concepts—cover cropping and green manuring—go well together, as most cover crops are easily used as green manures prior to the planting of a commercial crop. The combined benefits become economically feasible when the cover is grown during the off-season or inter-seeded with the main crop.

It is made even more desirable when the cover crop includes nitrogen-fixing legumes.

### Manuring and Composting

Livestock manures are the most traditional and widely recognised organic fertilisers. Under ideal circumstances, livestock enterprises are integrated into the whole farm operation, and manuring becomes part of a closed system of nutrient recycling. This is still strongly encouraged in organic operations. In reality, however, crops and livestock production are often divorced from each other, and manures must be imported. This has created some concerns in the organic community, since much manure is now generated by large, industrial agriculture feeding operations called CAFOs (Confined Animal Feeding Operations). Not only are there concerns about contaminants (heavy metals, antibiotics, pesticides, hormones) but many in the organic community also object to any "partnering" with this segment of conventional agriculture, which is considered at odds with the environmental and social values represented by organic farming.

Nonetheless, the National Organic Program does not differentiate between CAFO and other livestock manure sources. However, the NOP regulations do require that livestock manure not contain any synthetic substances not included on the National List of synthetic substances allowed for use in organic crop production. Another issue that has grown up around manure use in organic farming relates to food safety. At a time when concerns about microbial contamination are high, there are questions about the risks associated with manure use on food crops.

A focus piece on the February 2000 television news programme 20/20 was especially controversial. The segment suggested that organic foods were more dangerous than other food products in the marketplace due to manure fertilisation. The reporter ignored the fact that conventional farms also use manures. Were all the manure generated annually in the U.S. (about 1.4 billion tons) applied only to organic farm acreage (estimated at roughly 1.5 million acres in 1997), each acre would receive about 933 tons.

Furthermore, certified organic producers have strict guidelines to follow in handling and applying manures. The National Organic Program regulations require raw animal manure be incorporated into the soil not less than 120 days prior to the harvest of a product whose edible portion has direct contact with the soil surface or soil particles, and be incorporated into the soil not less than 90 days prior to harvest of a product whose edible portion does not have direct contact with the soil surface or soil particles.

One of the best means of handling manures is composting. Composting stabilises the nutrients in manure, builds populations of beneficial organisms, and has a highly beneficial effect on soils and crops. Compost can be produced on-farm by a number of means. Additional products from composts, such as compost teas, have special applications in organic agriculture. Human manures are expressly forbidden in certified organic production. This includes composted sewage sludge (also called "biosolids").

The organic community made its opinion on this quite clear when the USDA's first draft of the national rule proposed allowing the use of sludge in certified production. It was counted as one of the "big three" targets of protest, along with food irradiation and genetic engineering. The prohibition of biosolids would have been disconcerting to Albert Howard, who decried the failure of cities to return their organic wastes to the countryside. Such recycling was, in his mind, a key aspect of sustainability.

***Livestock on Organic Farms***

Among the thorniest of issues swirling around the edges of organic agriculture is the role of livestock. The disagreements arise because of the diversity of people and philosophies in the organic community. Organic agriculture can usually count vegetarians and animal welfare proponents among its more vocal supporters. Many of these people feel strongly that animals should not be exploited. Their rationale often goes beyond emotional and religious beliefs; convincing human health concerns, social issues, and environmental reasons are commonly cited. On the other side of this argument are those who feel that an organic farm cannot achieve its full potential or ecological balance without livestock manure; that it is essential to nutrient cycling and to the finer aspects of soil building.

Excellent soil fertility can be built in the absence of farm livestock and livestock manures by using vegetation-based composts and by harnessing the livestock in the soil—earthworms and other soil organisms. However, it is clearly easier to design a contemporary, low-input organic farm when traditional livestock are integrated. The biological and enterprise diversity that livestock can bring contributes enormously to stability and sustainability. A good example is provided by Rivendell Gardens in Arkansas, which began integrating livestock enterprises after several years as a solely horticultural operation.

The owners of Rivendell, Gordon and Susan Watkins, now rotate their strawberry and vegetable crops with grass-fed, direct-marketed beef and pastured poultry. Ideally, poultry follows beef on pasture to reduce cattle parasites. The seasons in mixed legume/grass pasture leave the soil quite mellow and well-manured for subsequent high-dollar horticultural crops. The Rivendell operation demonstrates the sort of organic management where a large number of organic farmers and many animal welfare proponents find common ground. The Watkins' animals are all raised with minimal confinement and generous access to sunshine, fresh air, and free-choice foodstuffs.

While domesticated and destined for slaughter, they lead low-stress lives in conditions much closer to natural than the conventional alternatives. This is the antithesis of industrialised factory farming systems, which are increasingly becoming the norm in livestock production.Many in both the organic and animal welfare communities are working to prohibit factory farming of livestock in organic systems. Many of the difficulties revolve around the fine interpretations of language in various organic standards. Wording such as "access to fresh air and sunlight," for example, can be construed to mean nothing more than opening the door on one end of a large confinement poultry house for a couple hours a day.

What Howard had not taken into account is the almost universal contamination of urban wastes with heavy metals and chemicals that are not eliminated by composting and may even be concentrated. Perhaps this was not yet a serious problem in his time; it is, however, in ours. Organic farmers and consumers concerned about contamination of soil and crops with agricultural pesticides and synthetic fertilisers would be remiss to ignore the contamination hazards of even well-composted sewage.

#### *Intercropping and Companion Planting*

Interplanting two or more mutually beneficial crops in close proximity is one strategy for increasing biodiversity. In large-scale mechanised crop culture, this is called intercropping. It typically involves alternating rows or a number of rows of compatible field crops, like soybeans and corn. It also applies to sowing multiple forage crops, like alfalfa, bromegrass, and timothy, when these are grown together. When interplanting is done on a smaller scale, it is often called companion planting. A classic example of companion planting is the inter-planting of corn with pole beans and vining squash or pumpkins. In this system, the beans provide nitrogen; the corn provides support for the beans and a "screen" against squash vine borer; the vining squash provides a weed suppressive canopy and discourages corn-hungry raccoons.

#### *Biological Pest Control*

Organic farming relies heavily on populations of beneficial insect predators and parasites, pest disease agents, insect-eating birds and bats, and other creatures, to help manage pest problems. These biological controls help keep pest numbers at levels where further cultural activities or relatively mild pesticides are (usually) adequate to assure a crop. In some instances, biological control can be so effective that no additional action is even needed by the farmer. Some see biological control as a default benefit of the soil fertility practices of organic farming.

The diversity of crops in a soil-building rotation, the use of cover crops, and other practices build a diverse soil biology that works to keep soil pests in check. They also provide substantial above-ground habitat for beneficials. The absence of pesticides also favors biocontrol. In many organic systems, farmers sometimes purchase and release control agents like ladybird beetles, lacewings, trichogramma wasps, etc., or use weeder geese – a quaint but effective biological weed control. Increasingly, growers are designing and maintaining both permanent and temporary habitats specifically for beneficial insects, spiders, and other helpful species. This is known as farmscaping.

#### Sanitation

Sanitation can take on many forms:

- removal, burning, or deep plowing of crop residues that could carry plant disease or insect pest agents
- destruction of nearby weedy habitats that shelter pests
- cleaning accumulated weed seeds from farm equipment before entering a new, "clean" field
- sterilising pruning tools

As in human and animal health, sanitation practices can go a long way in preventing crop pest problems. However, many practices – such as clean cultivation, deep plowing, and burning crop residues – can increase erosion and reduce biodiversity. Thus, they may conflict with sustainability. Good organic growers recognise this and treat those practices as transitional or rescue options, rather than relying on them on an annual basis.

#### Tillage and Cultivation

Tillage and cultivation are tools that can accomplish a variety of objectives in farming systems: weed control, crop residue management, soil aeration, conservation of manures and other fertilisers, hardpan reduction, sanitation to destroy pest and disease habitat, etc. While conventional farmers rely on chemicals to accomplish many of these objectives, organic growers focus more on improving tillage and maximising its benefits. Guidelines for primary tillage, for example, are intent on conserving crop residues and added manures in the upper, biologically active zones of the soil, rather than burying them deeply where decomposition is anaerobic (oxygen-starved).

Leaving soils completely bare and vulnerable to erosion is discouraged; fall moldboard plowing is certainly frowned upon. Cultivation in organic systems often rises to the level of art. Row-crop farmers frequently use blind cultivation -shallow tillage, which largely ignores the crop rows-beginning shortly after seeding until the plants are but a few inches high. Rotary hoes, wire-tooth harrows, and similar equipment can be used for blind cultivation, delaying the first flush of weeds and giving the crop a head start.

Organic agriculture is often characterised as addicted to maximum tillage – with growers using every opportunity to lay the land bare with shovel, plow, or rototiller. This image has been magnified through the popularity of small-scale organic systems like the French Intensive and Biointensive Mini Farming models that espouse double- and triple-digging to create deep rooting beds. While appropriate to such intensive systems, this degree of cultivation is not characteristic of organic agriculture in general.

It may surprise some to learn that a large number of organic producers are not only interested in conservation tillage, they have adopted it. This will be a surprise because many believe that conservation tillage always requires herbicides.The interest in conservation tillage among organic producers in the Cornbelt was well documented in the mid-1970s by Washington University researchers. They noted that the vast majority of organic farmers participating in their studies had abandoned the moldboard plow for chisel plows.

Plowing with a chisel implement is a form of mulch tillage, in which residues are mixed in the upper layers of the soil, and a significant percentage remains on the soil surface to reduce erosion. Furthermore, a notable number of organic farmers had gone further to adopt ridge-tillage, a system with even greater potential to reduce erosion. It was especially interesting to note that the use of these conservation technologies was almost nil among neighboring conventional farms at the time. Organic growers were actually pioneers of conservation tillage in their communities.

Determining the amount, the timing, and the kind of tillage to be done can be a balancing act for the organic grower, but experience and observation over time lead to proficiency. There are downsides to tillage, however, and most organic growers are well aware of them. The most obvious of these is the dollar cost; organic farmers are as concerned as their conventional counterparts about costs of production and strive to minimise expensive field operations. There is also a cost to the soil and environment. Every tillage operation aerates the soil and speeds the decomposition of the organic fraction. While this may provide a boost to the current crop, it can be overdone and "burn up" the humus reserves in the soil. Excessive tillage can also be directly destructive to earthworms and their tunneling, reducing their benefits to the land. There is also the danger of compaction, even when field operations are well timed.

#### Mulching

Mulching is a practice often used by organic growers.

Traditionally, it entails the spreading of large amounts of organic materials—straw, old hay, wood chips, etc.—over otherwise bare soil between and among crop plants. Organic mulches regulate soil moisture and temperature, suppress weeds, and provide organic matter to the soil. Mulching is most appropriate to small, intensive operations with high-value annual or fruit crops. A few systems of no-till organic gardening have evolved from the concept of deep, permanent mulching. Among these are the well-known Ruth Stout method and Synergistic Agriculture—a raised bed system developed by Emilia Hazelip, who adapted concepts from Permaculture and the ideas of Masanobu Fukuoka.

Mark Cain and Michael Crane, co-owners of Dripping Springs Gardens—an intensive market gardening operation in Arkansas—have adapted Emilia's system to their farm with considerable success. Plastic mulch, as long as it is removed at end of growing or harvest season, is also permitted in certified organic production. Its use allows larger acreage to be brought more easily under herbicide-free management, though there are serious issues to be addressed.

#### High-Input Organic Agriculture

Organic farming is a system that uses a minimum of off-farm inputs. In these systems, traditional rotations and soil building practices are usually employed, followed by clean cultivation and the laying of plastic mulch and drip irrigation tape on shaped beds. During the season, large amounts of soluble organic fertilisers—typically fish-based—are fed to the crop through the drip system (i.e., organic fertigation). At the end of the season, all plastics must be removed from the field, and it is returned to more standard organic management. Ideally, an off-season cover crop will be planted. Such systems are often exceptionally productive and economically attractive, when organic premiums are good.

The high cost of soluble organic fertiliser (typically hundreds of dollars/acre), however, plus the marginally higher cost of pest controls, make such systems largely non-competitive in the conventional marketplace.The labeling of such high-input systems as organic presents a paradox for many proponents of organic agriculture. It is unclear whether these technological advancements reflect the kind of farming most practitioners and supporters of organics think of as truly "organic." To begin with, the research citing environmental and economic benefits has largely been done on low-input organic systems; it is questionable whether similar findings would be made about high-input systems, especially regarding environmental matters.

Of particular note, while low-input organic systems are documented as being more resistant to erosion, fields under plastic mulch are reported to be fifteen times more erodible. Traditional organic farms leach minimal amounts of nitrogen into tile or groundwater; the losses from fields loaded with high levels of soluble organic fertilisers is certain to be greater, but how much greater is unknown.

The fossil fuel energy involved in plastic manufacture, transportation, and application may or may not be compensated by reductions in tractor fuel use. Finaliy, the lowered capital investment required to produce a crop by traditional organic methods makes this form of farming more accessible to resource-poor farmers and entails less risk in years of crop failure or lack of premiums. These factors are less certain in a high-input system. A further consideration is the issue of plastic disposal following removal. At this time, there are few to no options for recycling, and landfills are the fate of plastic mulches at the end of each season.While it is unwise to rush to judgement regarding high-input organic farming, it is clear that some adaptations will need to be made, if the traditional character and sustainability benefits of organic farming are to be preserved.

### *Fire*

While fire can be used in a number of ways in organic agriculture, the area of greatest interest is flame or thermal weeding. In its most common application, torches mounted on a tractor toolbar throw a hot flame at the base of mature (i.e., heat-resistant) plants, over the inter-row area, or both. Tractor speed is adjusted so that weeds are not burned so much as seared. Searing is sufficient to kill most seedling weeds and uses less fuel. Liquid propane (LP) gas is the fuel most commonly used, though alternatives such as alcohol and methane offer the possibility of on-farm sources.

### *Supplemental Fertilisation*

In many organic systems, crop rotation, manuring, green manuring, along with enhanced biological activity in the soil, provide an abundant supply of plant-essential minerals annually. This is especially true on naturally deep and rich prairie soils. It is less true on poorer soils and on those that have been heavily exploited through non-sustainable farming practices. To correct mineral deficiencies in organically managed soils, organic growers often apply ground or powdered rock minerals. The most commonly used rock mineral is high-calcium aglime. Dolomitic limestone, various rock phosphates, gypsum, sulfate of potash-magnesia, and mined potassium sulfate are also common. These are all significant sources of primary (P, K) and/or secondary (Ca, Mg, S) plant nutrients. The savvy organic grower applies significant amounts of these materials only with the guidance of regular soil testing.

Less common are other rock powders and fines that are limited sources for the major nutrients but are rich in micronutrients or have some other soil-improving characteristic. Among these are glauconite (greensand), glacial gravel dust, lava sand, Azomite®, granite meal, and others. Supplementary nutrients that include nitrogen are often provided in the form of animal or plant products and by-products such as fish emulsion, blood meal, feather meal, bone meal, alfalfa meal, and soybean meal. Most of these products also supply some organic matter, though that is not the primary reason they are used.

One basis on which to evaluate the tools or practices one chooses for an organic operation is whether or not they contribute to biodiversity—biodiversity being one of the principal characteristics of a sustainable organic agriculture.Crop rotation, cover cropping, farmscaping, companion planting, and intercropping are outstanding examples of practices that contribute to biodiversity. They therefore contribute to the long-term stability and sustainability of the farm agroecosystem.

Composting and manuring likewise contribute to biodiversity, but since the diversity they promote is mostly in soil biota, it is rather less obvious to the casual observer. On the other end of the spectrum are practices such as tillage, cultivation, thermal weeding, solarisation, and plastic mulching. Such tools significantly reduce diversity in the field and tend to move the system in a less sustainable direction. This, however, does not necessarily make these practices bad choices.Organic farming has often been called natural farming, as it tries to mimic the processes of nature in producing crops and livestock.

However, the analogy goes only so far, since most agricultural systems are characterised by a struggle between human and nature, each with a clear notion of what plants and animals the land ought to support and in what proportions. The farmer is typically in the position of "holding back" the natural succession of plant and animal species through the use of diversity-reducing tools and practices. The ideal is to bring about an agricultural system in which the long-term direction emphasises diversity and sustainability. Among the best visualisations of this ideal are those emerging from the Permaculture movement.

#### Balanced Nutrition

Whether or not it has been customary in the past, organic growers are encouraged to have periodic soil testing done on their fields. How the results of a soil test are used, however, can vary considerably among farmers, depending on their personal philosophy and management skill.While there is certainly a segment of the organic farming community that has no faith in soil audits, most growers use them as a means to monitor progress in building their soils, to identify nutrient deficits, and to guide supplementary fertilisation.

While there are many ideas about fertilisation guidelines, there are two schools of thought that dominate. The first is commonly known as the sufficiency approach or model. Though somewhat oversimplified, the following are among its significant characteristics:

- annual fertiliser additions of P and K are based in good part on how much the crop is expected to remove from the soil at harvest
- additional amounts of P and K are recommended based on keeping the soil nutrient reserves at a particular level

- lime is added to the soil based on pH
- little to no attention is paid to nutrient balance or to the levels of secondary nutrients Ca and Mg

The second approach is referred to variously as cation nutrient balancing, the Albrecht system, the CEC, or the base saturation approach. It differs from the sufficiency approach in that fertiliser and lime recommendations are made based on an idealised ratio of nutrients in the soil and its capacity to hold those nutrients against leaching.Cation nutrient balancing is more popular among practitioners of organic farming and sustainable agriculture in general than it is among conventional growers. However, there is no universal agreement on which approach is most appropriate for organic management.

A wide range of other products -humates, humic acids, enzymes, catalyst waters, bioactivators, surfactants, to name a few—are also acceptable in organic crop production. However, the tradeoff between out-of-pocket cost and efficacy of such materials is often challenged by conventional and organic growers alike. Organic growers are encouraged to experiment, but to do so in a manner that allows the actual results to be measured.

### Biorational Pesticides

While, in principle, any pesticide use is discouraged in organic systems, a rather wide range of biorational pesticides is permitted. The frequency of pesticide use varies considerably with crop and location. For example, there is virtually no use of pesticides on organic row crops in the Cornbelt. By contrast, organic tree fruits in the Midsouth routinely receive heavy applications of several fungicides and insecticides allowable in organic production. The pesticides permitted in organic farming fall predominantly into several classes.

*Minerals*: These include sulfur, copper, diatomaceous earth, and clay-based materials like Surround®.

*Botanicals*: Botanicals include common commercial materials such as rotenone, neem, and pyrethrum. Less common botanicals include quassia, equisetum, and ryania. Tobacco products like Black-Leaf 40® and strichnine are also botanicals but are prohibited in organic production due to their high toxicity.

*Soaps*: A number of commercial soap-based products are effective as insecticides, herbicides, fungicides, and algicides. Detergent-based products are not allowed for crop use in organic production.

*Pheromones*: Pheromones can be used as a means to confuse and disrupt pests during their mating cycles, or to draw them into traps.

*Biologicals*: One of the fastest-growing areas in pesticide development, biopesticides present some of the greatest hope for organic control of highly destructive pests. Among the most well-known biopesticides are the Bacillus thuringiensis (Bt) formulations for control of lepidopterous pests and Colorado potato beetle.

**Problems in Organic Agriculture**

One of the challenges faced by organic growers is getting suitable planting stock. Certified production requires that seed not be treated with pesticides and be organically grown.Transplants must be purchased from a certified organic source or otherwise be grown using organic methods. This presents some difficulty, as most commercial potting mixes contain prohibited fertilisers and wetting agents, requiring special ordering or the added effort of making a homemade mix.

Generally, planting stock for perennial crops like tree fruits and berries is available from conventional sources, even if treated with pesticides. However, no production can be sold as organic for a minimum of 12 months following transplanting to an organic field. In all instances, organic growers are not permitted to use varieties of crops that have been developed through genetic engineering (GE). At first glance, this might seem perplexing. GE crops promise further means of non-chemical pest control and the possibility of nutritionally enhanced foods. However, the organic community is concerned about environmental, economic, and social impacts from this new technology which, they feel, have not been adequately studied. This same concern applies to GE foods.

Natural foods consumers—a large segment of the organic market—do not want them. This was made abundantly clear to the USDA when its first draft of the proposed rule, which suggested permitting genetically modified organisms (GMOs), was released in 1997. The backlash was so strong that not only GE crop varieties, but all inputs, such as GMO-derived biopesticides, are prohibited from organic production.Unfortunately the GE issue does not resolve itself so simply for organic farmers. The proliferation of GMOs in the marketplace and across the landscape has created a host of new challenges that have particular implications for organic agriculture:

– *Genetic drift onto crops.* Pollen drifting from adjacent fields can contaminate organic crops. Some clients in the lucrative European Union market accept 0% GMO contamination; others accept only very low levels. Pollen from cross-pollinating crops like corn easily travels hundreds of yards and may be carried for miles under the right conditions. No reasonable amount of non-crop buffer can prevent contamination except in highly isolated locations.

- *Genetic drift onto soils.* Concerns have been expressed over the persistence of GE pollen in soils. Should contaminated soils be decertified?
- *Non-labeled products.* Should soybean, cottonseed, and corn gluten meals be prohibited as fertilisers, since these may come from GE varieties? Should pesticidal use of vegetable oils be prohibited?
- *Pest control problems.* The rapid proliferation of corn, cotton, and other crops featuring the Bt gene from Bacillus thuringiensis is likely to accelerate the development of resistance to this natural pesticide. Organic growers may lose one of their most useful pest management tools.
- *Domination of the food system.* In the late 1990s, the public was made aware of two specific developments of agricultural biotechnology dubbed the terminator and traitor technologies.

Terminator technology is the genetic altering of seed so that the subsequent generation will not germinate. This prevents farmers from saving a portion of the harvested crop to replant the next year—a traditional practice among farmers worldwide. Traitor or suicide technology is the genetic altering of seed so that it will fail to germinate or mature unless a proprietary chemical is applied. Such technologies are authoritarian and advance the centralisation and industrialisation of the food system. Organic farmers perceive additional challenges from genetic drift impacts and difficulty in finding non-GE seed in the marketplace.

### Foliar Fertilisation

Foliar fertilisation or feeding entails the application—via spraying—of nutrients to plant leaves and stems and their absorption at those sites. It is not specifically an organic practice, though it is commonly used by many organic growers. The fertiliser materials used are typically soluble fish- and seaweed-based products, naturally chelated nutrients, humic acid extracts, and teas made from plants, dried blood, manure, guano, or compost. At first glance, the use of foliar feeding appears contradictory to the organic notion that one feeds the soil to feed the plant. Organic growers rationalise the use of this approach on two points.

1. Foliar feeding is strictly supplemental fertilisation; it is not used as a substitute for traditional soil building practices.
2. Foliar fertilisation is understood to increase the production of root exudates, which stimulates biological activity in the rhizosphere (soil area adjacent to plant roots). The soil bio-life gets considerable benefit in this indirect way from foliar feeding.

There are a number of farming and gardening practices based on belief in a non-physical world closely aligned with our physical reality. Those who use these practices believe that conditions in this unseen realm influence, or even dictate, what happens on a material level. They take pains to understand these influences and adjust their farm activities. In some instances, they act to influence events. Obviously, these practices are rooted more deeply in metaphysics than in the conventional sciences. The most well-known of these esoteric practices is the scheduling of planting or other field operations according to lunar or other astrological signs.

Dowsing and other forms of divining may also be used to guide scheduling, fertiliser selection, and other facets of farming. The use of potentised preparations, as done in Biodynamic™ farming, is also a metaphysical practice. Radionics, energy balancing towers, "medicine" wheels, prayer—these are further examples of practices used by some farmers to bring a tangible spiritual element into their farm operation. Esoteric practices like these (there are many others) are not inherently part of organic agriculture. In all likelihood, only a small-to-modest minority practice them or believe in their efficacy. Still, esoteric practices are sometimes associated with organic agriculture, and it is almost a given that practitioners of "energetic agriculture" are either certified organic or actively pursue a sustainable form of agriculture.

The rationale for this association of organic with the esoteric is not difficult to understand. Most metaphysical practices are founded on a world-view similar to that of ecology, in which everything is related to everything else. The difference is that ecology considers this interrelationship only on a hard physical level; metaphysics perceives it also on a non-physical level. This belief tends to cast humankind as part of nature, not merely as an inhabitant. This perspective would tend to make one more concerned about the environment and, therefore, more readily drawn to organics as a philosophy and pragmatic farming approach.

To put it in a nutshell, it is natural for the metaphysically-inclined to choose organics; it is far less likely, however, that the average organic farmer is drawn to metaphysical practices. A further factor plays into the association between esoteric farming practices and organics. It relates to the fact that organic agriculture has, for many years, been marginalised and treated as pseudoscience by the mainstream—an experience shared by practitioners of metaphysical arts. It is not surprising for underdogs to seek common ground.

***Buffers and Barriers***

Field buffers are strips of sod or other permanent vegetation at the edge of or surrounding crop fields. There is considerable interest in both conventional and alternative agriculture in these buffers, since they assist in reducing soil erosion and improving water quality. If managed properly, some buffers also serve as beneficial insect habitat. In certified organic production, however, field buffers have an added purpose in reducing crop contamination from chemicals used on adjacent land.

Most agencies require a minimum 25-foot buffer along "uncontrolled" borders where there is a hazard of chemical use. Wider borders may be required where hazards are great—for instance, where adjacent farms have synthetic pesticides applied by aircraft.

The pioneers of organic farming considered organics the referred direction for the whole of agriculture to take. It is likely that most contemporary proponents still hold that view. While recent growth in the organic industry is definitely encouraging, much of the impetus is tied to its growth as a niche market, not as a serious shift in the direction of mainstream agriculture. Unfortunately, a likely reason for the newfound "tolerance" of organic agriculture in many land grant universities and other formerly hostile environs is its perception as a niche market opportunity only.

As such, it is not a serious threat to the status quo.Is it practical and responsible to promote organic agriculture as the dominant approach to farming in the future? While some are quick to answer, "Yes!", most are buffaloed by the nagging question, "Can organic agriculture feed the world?" Questions about the productivity and the prospects of widespread starvation have long been an effective tactic for stonewalling serious discussion of organics. There are a few points that can and should be made regarding the "starving billions" scenario.

Critics warn that, were organic farming to be adopted on a wider scale, per-acre agricultural productivity would decline sharply; that meeting food needs would necessitate the plowing of even more erodible hillsides and the draining of more wetlands. Such scenarios are often based on pre-chemical era yield data that ignore the advancements of modern organic farming. Contemporary research on organic systems—as limited as it is—indicates that the per-acre productivity of organic and conventional systems is not vastly different.

Other scenarios are sometimes based on the perceived need for vast quantities of grain—particularly corn. Since overproduction of grain has led to prices below costs of production for several years, this is a questionable basis for argument. It also ignores the fact that most of the grain produced is fed to livestock and that many livestock species can eat forage instead. Ruminant livestock—cattle and sheep especially—are designed by nature to thrive on forage and can even be finished for market on pasture. The paucity of good research on the productivity of organic systems also spotlights the lack of practical research for organic producers.

That organic farms are as productive as they appear to be is remarkable in light of minimal research and extension support for many decades. It has often been argued that, had comparable resources been put into organic research, its widespread feasibility would be unquestionable. This assertion is well-supported by the development, in recent years, of a number of alternative, organically acceptable pesticides, that are certain to expand organic production of many crops in regions where it had previously been near impossible on a commercial basis.

Increasing agricultural production alone does not alleviate hunger. The amount of grain produced in the world in 1999 could, by itself, sustain 8 billion people—2 billion more than our current population. Total food production is estimated to provide each human being with at least 3500 calories per day. The issue of hunger, it appears, is not so much a lack of food, but lack of entitlement to food. People are shocked to learn that while many Bangladeshees starved during the 1974 floods, roughly 4 million tons of rice produced in that country were stacked in warehouses for want of buyers.

The people were simply too poor to buy it. They are surprised to learn that Ireland exported grain during the Irish potato famine while 1 million of its citizens died and even more emigrated; that India regularly exports food and animal feed despite an estimated 200 million in starvation. Those lacking the ready cash to buy food or the resources to produce it themselves seem destined for hunger no matter what miracles agricultural technology provides. The world's nations will need to deal with issues of equity and democracy first, if hunger is ever to be effectively addressed.

No agriculture can continue to feed a growing population if it depletes or fouls its resource base. The path undertaken by conventional agriculture is ultimately a dead end in this regard, though there is an almost mystical faith that genetic engineering and other complex technologies will always triumph. Agriculture needs to be sustainable. Therefore, those who promote organic agriculture as a true alternative are well advised to do their part in ensuring that certification and regulation does not create a "compliance agriculture" in which sustainability becomes little more than an afterthought.

Few would consider record keeping as a tool for organic crop production. However, the documentation of how and where a crop was raised, what products were applied and when, which bin or cooler it was stored in, is of critical importance to establishing the integrity of the product. If the farmer cannot provide reasonable documentation that his or her crop was organically grown, that it has not been contaminated with chemicals, and that it has not commingled with a similar conventional product, then certification may well be denied.

As stated at the outset, the tools and practices described in detail here are by no means an exhaustive listing of organic options. Among those not discussed but requiring mention are timing of planting and field operations to avoid pests or disrupt their life cycles, various forms of pest traps, physical barriers to pests, and increasing plant populations to enhance crop-weed competition.

# Bibliography


Anderson D C., "Use of cereal residues in beef cattle production systems", *Journal of Animal Science,* 1978.

Conner M C and Richardson C R., "Utilization of cotton plant residues by ruminants", *Journal of Animal Science,* 1987.

Klopfenstein T. Roth L, Rivera S F and Lewis M., "Corn residues in beef production systems", *Journal of Animal Science,* 1987.

Males J R., "Optimizing the utilization of cereal crop residues for beef cattle", *Journal of Animal Science,* 1987. 65:1124-1130.

McDowell R E and Hildebrand P E., "Integrated crop and animal production: Making the most of resources available to small farms in developing countries", *Working papers*: The Rockefeller Foundation, New York, NY, USA, 1980.

Owen E., "Crop residues as animal feeds in developing countries - use and potential use", In: M Wanapat and C Devendra (eds), *Relevance of crop residues in animal feeds in developing countries,* Funny Press, Bangkok, 1985.

Tilley J M A and Terry R A., "A two-stage technique for the in vitro digestion of forage crops", *Journal of the British Grassland Society,* 1963.

Timon V M and Hanrahan J P (eds), "Small ruminant production in the developing countries", *FAO Animal Production and Health Paper* 58. Food and Agriculture Organization of the United Nations, Rome, Italy, 1986.

Wahed R A., "Stall-feeding barley straw to goats: The effect of refusal-rate allowance on voluntary intake and selection", *Ph.D. thesis,* University of Reading, Reading, UK, 1987.

Wahed R A and Owen E., "Comparison of sheep and goats under stall-feeding conditions: Roughage intake and selection", *Animal Production,* 1986.

Wahed R A and Owen E., "The effect of amount offered on selection and intake of barley straw by goats", *Animal Production,* 1986.

World Bank and Winrock International, "Sheep and goats in developing countries: Their present and potential role", World Bank and Winrock International, Washington, DC, USA, 1983.

# Index